Holt Mathematics

Chapter 2 Resource Book

HOLT, RINEHART AND WINSTON
A Harcourt Education Company
Orlando • Austin • New York • San Diego • London

Copyright © by Holt, Rinehart and Winston

All rights reserved. No part of this publication may be reproduced or transmitted in any form or by any means, electronic or mechanical, including photocopy, recording, or any information storage and retrieval system, without permission in writing from the publisher.

Teachers using HOLT MATHEMATICS may photocopy complete pages in sufficient quantities for classroom use only and not for resale.

Printed in the United States of America

If you have received these materials as examination copies free of charge, Holt, Rinehart and Winston retains title to the materials and they may not be resold. Resale of examination copies is strictly prohibited and is illegal.

Possession of this publication in print format does not entitle users to convert this publication, or any portion of it, into electronic format.

ISBN 0-03-078192-2

7 8 9 10   170   09

CONTENTS

Blackline Masters

Parent Letter	1
Lesson 2-1 Practice A, B, C	3
Lesson 2-1 Reteach	6
Lesson 2-1 Challenge	7
Lesson 2-1 Problem Solving	8
Lesson 2-1 Reading Strategies	9
Lesson 2-1 Puzzles, Twisters & Teasers	10
Lesson 2-2 Practice A, B, C	11
Lesson 2-2 Reteach	14
Lesson 2-2 Challenge	15
Lesson 2-2 Problem Solving	16
Lesson 2-2 Reading Strategies	17
Lesson 2-2 Puzzles, Twisters & Teasers	18
Lesson 2-3 Practice A, B, C	19
Lesson 2-3 Reteach	22
Lesson 2-3 Challenge	23
Lesson 2-3 Problem Solving	24
Lesson 2-3 Reading Strategies	25
Lesson 2-3 Puzzles, Twisters & Teasers	26
Lesson 2-4 Practice A, B, C	27
Lesson 2-4 Reteach	30
Lesson 2-4 Challenge	31
Lesson 2-4 Problem Solving	32
Lesson 2-4 Reading Strategies	33
Lesson 2-4 Puzzles, Twisters, & Teasers	34
Lesson 2-5 Practice A, B, C	35
Lesson 2-5 Reteach	38
Lesson 2-5 Challenge	39
Lesson 2-5 Problem Solving	40
Lesson 2-5 Reading Strategies	41
Lesson 2-5 Puzzles, Twisters & Teasers	42
Lesson 2-6 Practice A, B, C	43
Lesson 2-6 Reteach	46
Lesson 2-6 Challenge	47
Lesson 2-6 Problem Solving	48
Lesson 2-6 Reading Strategies	49
Lesson 2-6 Puzzles, Twisters & Teasers	50
Lesson 2-7 Practice A, B, C	51
Lesson 2-7 Reteach	54
Lesson 2-7 Challenge	55
Lesson 2-7 Problem Solving	56
Lesson 2-7 Reading Strategies	57
Lesson 2-7 Puzzles, Twisters & Teasers	58
Lesson 2-8 Practice A, B, C	59
Lesson 2-8 Reteach	62
Lesson 2-8 Challenge	63
Lesson 2-8 Problem Solving	64
Lesson 2-8 Reading Strategies	65
Lesson 2-8 Puzzles, Twisters, and Teasers	66
Algebra Tiles	67
Cards for Reaching All Learners	68
Answers to Blackline Masters	69

Date _____

Dear Family,

In this chapter students will be introduced to fundamental algebraic concepts such as variables, algebraic expressions, and inequalities. Students learn to write expressions for data in tables, to solve algebraic equations by using inverse operations, and to solve word problems by writing algebraic equations.

A **variable** is a letter or symbol that represents a quantity than can change.

An **algebraic expression** contains one or more **variables.**

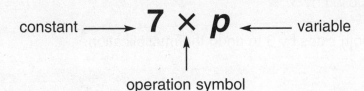

To evaluate an **algebraic expression,** you substitute a number for the variable and then find the value. For example if $p = 4$, the value of the above expression would be 28.

An **equation** is a mathematical statement that says two quantities are equal. You can compare an equation to a balanced scale. An equation is solved by finding a **value** for the variable that makes the equation true.

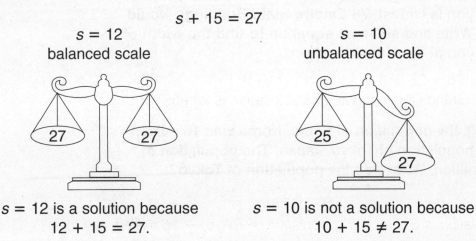

An **inequality** is a statement that two quantities are not equal.

$$15 < 18 \qquad 12 \leq 29 \qquad 83 > 17$$

Addition and subtraction are **inverse,** or opposite, operations; so are multiplication and division. **Inverse** operations can be used to solve various kinds of equations.

Holt Mathematics

If an equation involves addition, we solve it by subtracting from both sides. This "undoes" the addition.

$$x + 62 = 93$$
$$\underline{-62 \quad -62} \quad \text{Subtract 62 from both sides to undo the addition.}$$
$$x + 0 = 31 \quad x = 31$$

When an equation contains multiplication, we use division to undo the multiplication.

$$3x = 12$$
$$3x = 12 \quad x \text{ is multiplied by 3.}$$
$$\frac{3x}{3} = \frac{12}{3} \quad \text{Divide both sides by 3 to undo the multiplication.}$$
$$x = 4$$

Check: $3x = 12$

$3(4) \stackrel{?}{=} 12$ Substitute 4 for x in the equation.

$12 \stackrel{?}{=} 12$ ✓ 4 is the solution.

Knowing how to solve algebraic equations is a skill that students will use in other areas such as science, business, and social studies.

The Empire State Building is 381 m tall. At the location where the Grand Canyon is widest, 76 Empire State Buildings would fit end-to-end. Write and solve an equation to find the width of the Grand Canyon at its widest location.

Solution: 381 × 76 = 28,956 m
The width of the Grand Canyon at its widest location is 28,956 m.

In the year 2000, the population of Seoul, Korea was 16 million fewer than the population of Tokyo, Japan. The population of Seoul was 10 million. What was the population of Tokyo?

Solution:

10 million (population of Seoul) = T (population of Tokyo) − 16 million

10,000,000 = T − 16,000,000
26,000,000 = T

The population of Tokyo was 26,000,000.

For additional resources, visit go.hrw.com and enter the keyword MR7 Parent.

Name _____ Date _____ Class _____

LESSON 2-1 Practice A
Variables and Expressions

Circle the letter of the correct answer.

1. Which of the following is an algebraic expression?
   A 4 + 13
   B 10 • (3 − 2)
   C 15 ÷ 5
   D 9 − n

2. What is the variable in the expression (16 + a) • 5 − 4?
   F 16
   G a
   H 5
   J n

3. Which of these expressions is a way to rewrite the algebraic expression n ÷ 3?
   A $\frac{n}{3}$
   B n • 3
   C 3n
   D $\frac{3}{n}$

4. Which of these expressions is not a way to rewrite the algebraic expression n • 4?
   F n(4)
   G n • 4
   H $\frac{4}{n}$
   J 4n

Evaluate each expression to find the missing values in the tables.

5.
n	n + 3
1	4
2	
5	
7	
10	

6.
n	n • 2^2
2	8
3	
5	
7	
8	

7. If x = 3, what is the value of the expression 6 ÷ x?

   _____

8. If x = 2, what is the value of the expression 9 − x?

   _____

Name _____ Date _____ Class _____

LESSON 2-1 Practice B
Variables and Expressions

Evaluate each expression to find the missing values in the tables.

1.
n	n + 8²
7	71
9	
22	
35	

2.
n	25 − n
20	5
5	
18	
9	

3.
n	n • 7
8	56
9	
11	
12	

4.
n	24 ÷ n
2	12
6	
4	
8	

5.
n	n + 15
35	
5	
20	
85	

6.
n	n • 2³
7	
4	
10	
13	

7. A car is traveling at a speed of 55 miles per hour. You want to write an algebraic expression to show how far the car will travel in a certain number of hours. What will be your constant? your variable?

_____

_____

8. Shawn evaluated the algebraic expression x ÷ 4 for x = 12 and gave an answer of 8. What was his error? What is the correct answer?

_____

_____

Copyright © by Holt, Rinehart and Winston.
All rights reserved.

Holt Mathematics

Name _____ Date _____ Class _____

LESSON 2-1 Practice C
Variables and Expressions

Evaluate each expression to find the missing values in the tables.

1.

n	n ÷ 15
30	
75	
15	
105	

2.

n	$3n - 2^3$
3	
8	
10	
29	

3.

n	n + 17
34	
55	
26	
100	

4.

l	w	l × w
5	3	
6	3	
7	3	
8	3	

Evaluate each expression for the given value of the variable.

5. $5x + 2$ for $x = 4$

6. $63 - 8z$ for $z = 7$

7. $176 \div p$ for $p = 2$

_____ _____ _____

8. $\frac{64}{v} - 11$ for $v = 4$

9. $19w$ for $w = 5$

10. $98 - 5q$ for $q = 7$

_____ _____ _____

11. $48 \div n$ for $n = 3$

12. $x + x + x$ for $x = 15$

13. $16 + n^2$ for $n = 3$

_____ _____ _____

14. What is the next expression in the following pattern: $4n$; $8n$; $16n$?

15. What is the next expression in the pattern $x + 27$; $x + 24$; $x + 21$?

_____ _____

Copyright © by Holt, Rinehart and Winston.
All rights reserved.

Holt Mathematics

Name _____ Date _____ Class _____

LESSON 2-1 Reteach
Variables and Expressions

A variable is a letter or a symbol that stands for a number that can change. A constant is an amount that does not change.

A mathematical phrase that contains at least one variable is an algebraic expression. In the algebraic expression $x + 5$, x is a variable and 5 is a constant.

When you evaluate an algebraic expression, substitute a number for the variable and then find the value.

To evaluate the algebraic expression $m - 8$ for $m = 12$, first replace the variable m in the expression with 12.
$m - 8$
$12 - 8$
Then find the value of the expression.
$12 - 8 = 4$
The value of $m - 8$ is 4 when $m = 12$.

Evaluate each expression for the given value of the variable.

1. $x + 5$, for $x = 6$ 2. $3p$, for $p = 5$ 3. $z \div 4$, for $z = 24$ 4. $w - 7$, for $w = 15$

_____   _____   _____   _____

To find the missing values in a table, use the given values of the variable.

x	4x
3	12
4	■
5	■

Think: $x = 3$, so $4x = 4 \cdot 3 = 12$
Think: $x = 4$, so $4x = 4 \cdot 4 = 16$
Think: $x = 5$, so $4x = 4 \cdot 5 = 20$

Evaluate each expression to find the missing values in the tables.

5.

x	x + 7
3	10
5	
7	

6.

y	y − 2
9	
10	
14	

Name _____ Date _____ Class _____

LESSON 2-1 Challenge
Express Trains

Use the expression written on the side of each train's engine to find the missing values for the cars it pulls. Then choose your own value for the variable to fill in the last caboose on each train.

1.

Engine: $n \div 7$; Top: 6; Cars: $n = 42$, $n = 56$, $n = 28$

2.

Engine: $2x + 5$; Top: 11; Cars: $x = 3$, $x = 8$, $x = 10$

3.

Engine: $c \div 12$; Top: 4; Cars: $c = 48$, $c = 24$, $c = 60$

4.

Engine: $5p - 9$; Top: 31; Cars: $p = 8$, $p = 4$, $p = 11$

5.
Engine: $7m + 2m$; Top: 45; Cars: $m = 5$, $m = 2$, $m = 9$

Name _____ Class _____

LESSON 2-1 Problem Solving
Variables and Expressions

Write the correct answer.

1. To cook 4 cups of rice, you use 8 cups of water. To cook 10 cups of rice, you use 20 cups of water. Write an expression to show how many cups of water you should use if you want to cook c cups of rice. How many cups of water should you use to cook 5 cups of rice?

2. Sue earns the same amount of money for each hour that she tutors students in math. In 3 hours, she earns $27. In 8 hours, she earns $72. Write an expression to show how much money Sue earns working h hours. At this rate, how much money will Sue earn if she works 12 hours?

3. Bees are one of the fastest insects on Earth. They can fly 22 miles in 2 hours, and 55 miles in 5 hours. Write an expression to show how many miles a bee can fly in h hours. If a bee flies 4 hours at this speed, how many miles will it travel?

4. A friend asks you to think of a number, triple it, and then subtract 2. Write an algebraic expression using the variable x to describe your friend's directions. Then find the value of the expression if the number you think of is 5.

Circle the letter of the correct answer.

5. The ruble is the currency in Russia. In 2005, 1 United States dollar was worth 28 rubles. How many rubles were equivalent to 10 United States dollars?

    A 28
    B 38
    C 280
    D 2,800

6. The peso is the currency in Mexico. In 2005, 1 United States dollar was worth 10 pesos. How many pesos were equivalent to 5 United States dollars?

    F 1
    G 10
    H 15
    J 50

Name _____ Date _____ Class _____

Reading Strategies
LESSON 2-1 *Focus on Vocabulary*

The word **vary** means **change**. In math, a **variable** is a letter that holds a place for numbers that change.

1. Give some examples of things that vary.

   _____

   _____

   _____

The opposite of variable is **constant.** Something that is constant never changes, such as the street number of your house or the number of inches in a foot.

2. Give some examples of things that are constant.

   _____

   _____

In English, we use words in expressions such as, "see you soon" or "have a good day." In math, we use numbers and symbols to write **expressions** for other numbers.

$10 + 3$ $\qquad\qquad$ $4 + 8 + 5$ $\qquad\qquad$ $2(8 + 5)$

3. Write a math expression for 14.

   _____

4. Write a math expression for 25.

   _____

An **algebraic expression** is a math expression that contains a variable.

$x + 5$ $\qquad\qquad$ $3n + 1$ $\qquad\qquad$ $8 - w$

For Exercises 5–8, write "yes" if the expression is an algebraic expression or "no" if it is not.

5. $n + 7$ _____

6. $8(y + 1)$ _____

7. $6 + (10 + 5)$ _____

8. $4x - 1$ _____

Name _____ Date _____ Class _____

LESSON 2-1 Puzzles, Twisters & Teasers
Between Meals

What did the boat do after breakfast?

Circle each correct answer. Then put the letter above the correct answer in the box.

1. Which of the following is an algebraic expression?

L	C	N	D
$25x$	$0.25 \cdot 8$	$13 + 6 - 9$	$\frac{3}{4}$

2. Which of the following is NOT an algebraic expression?

T	M	C	A
$w + 7$	$15y$	$24 - g$	$250 - 135 \cdot 2$

3. In the following equation, which symbol is the variable?
   $x \cdot 2 = 6$

U	H	O	A	W
x	$\cdot$	2	$=$	6

4. Evaluate the expression for the given value of the variable.
   $x \cdot 9$ or $9x$ for $x = 3$

A	E	N
3	12	27

5. Solve the equation to find the value of the variable.
   $3n - 1 = 29$

S	C	P
9	10	11

6. Which algebraic equation best describes the following question?

   If a kangaroo can leap 6 feet, how many leaps will it take for her to travel 72 feet?

H	J	T
$6x = 72$	$6 + 72 = x$	$72x = 6$

Name _____ Date _____ Class _____

LESSON 2-2 Practice A
Translate Between Words and Math

Circle the letter of the correct answer.

1. Which of the following is the solution to an addition problem?
   A product
   B sum
   C plus
   D add

2. Which of the following is the solution to a subtraction problem?
   F minus
   G times
   H difference
   J less

3. Which word phrase represents the following expression: 5 • 3?
   A the product of 5 and 3
   B 5 less than 3
   C the quotient of 5 and 3
   D the sum of 5 and 3

4. Which word phrase represents the following expression: 14 ÷ n?
   F the difference of 14 and n
   G 14 more than n
   H take away n from 14
   J the quotient of 14 and n

Match each situation to its algebraic expression below.

A. 8 ÷ x   B. 8x   C. 8 − x   D. x + 8   E. x − 8   F. x ÷ 8

5. 8 take away x _____
7. the product of 8 and x _____
9. 8 more than x _____

6. x divided by 8 _____
8. the quotient of 8 and x _____
10. x decreased by 8 _____

11. Lily bought 14 beads and lost some of them. This situation is modeled by the expression 14 − x. What does x represent in the expression?

    _____

    _____

12. The pet store put the same number of hamsters in 6 cages. This situation is modeled by the expression 6n. What does n represent?

    _____

    _____

Copyright © by Holt, Rinehart and Winston.
All rights reserved.

Holt Mathematics

Name _____ Date _____ Class _____

LESSON 2-2
Practice B
Translate Between Words and Math

Write an expression.

1. Terry's essay has 9 more pages than Stacey's essay. If s represents the number of pages in Stacey's essay, write an expression for the number of pages in Terry's essay.

   _____

2. Let z represent the number of students in a class. Write an expression for the number of students in 3 equal groups.

   _____

Write each phrase as a numerical or algebraic expression.

3. 24 multiplied by 3   4. n multiplied by 14   5. w added to 64

   _____   _____   _____

6. the difference of 58 and 6   7. m subtracted from 100   8. the sum of 180 and 25

   _____   _____   _____

9. the product of 35 and x 10. the quotient of 63 and 9   11. 28 divided by p

   _____   _____   _____

Write two phrases for each expression.

12. $n + 91$ _____

13. $35 \div r$ _____

14. $20 - s$ _____

15. Charles is 3 years older than Paul. If y represents Paul's age, what expression represents Charles's age?

    _____

16. Maya bought some pizzas for $12 each. If p represents the number of pizzas she bought, what expression shows the total amount she spent?

    _____

Name _____ Date _____ Class _____

LESSON 2-2 Practice C
Translate Between Words and Math

Write each phrase as a numerical or algebraic expression.

1. the sum of 69, 140, and 300

2. 95 less than the quotient of x and 12

3. 144 less than 500

4. 22 added to the product of 14 and n

5. The difference of 98 and p, added to 4

6. 85 more than twice m

Write two phrases for each expression.

7. $\dfrac{150}{n}$

8. $79 - w$

9. $12 + 29q$

10. $(87 - p) + 11$

11. $(28 \div x) - 6$

12. $(4 + z) - 18z$

13. Mohamed bought several bottles of juice for $3 each. He paid for them all with a $20 bill. If j represents the number of bottles Mohamed bought, what expression represents the change he would receive?

14. A giant bamboo plant grew 18 inches per year. When Mrs. Sanchez started measuring the plant it was 5 inches tall. If y represents the number of years she measured the plant, what expression represents its height?

LESSON 2-2

Reteach
Translate Between Words and Math

There are key words that tell you which operations to use for mathematical expressions.

Addition (combine)	Subtraction (less)	Multiplication (put together groups of equal parts)	Division (separate into equal groups)
add plus sum total increased by more than	minus difference subtract less than decreased by take away	product times multiply	quotient divide

You can use key words to help you translate between word phrases and mathematical phrases.

A. 3 plus 5   **B.** 3 times x **C.** 5 less than p **D.** h divided by 6
    $3 + 5$ $3x$ $p - 5$ $h \div 6$

Write each phrase as a numerical or algebraic expression.

1. 4 less than 8
2. q divided by 3
3. f minus 6
4. d multiplied by 9

_____   _____   _____   _____

You can use key words to write word phrases for mathematical phrases.

A. $7k$
- the product of 7 and k
- 7 times k

B. $5 - 2$
- 5 minus 2
- 2 less than 5

Write a phrase for each expression.

5. $z \div 4$
6. $5 \cdot 6$
7. $m - 6$
8. $s + 3$

_____   _____   _____   _____

14   Holt Mathematics

Name _____ Date _____ Class _____

Challenge
LESSON 2-2
Animal State

Follow the steps below in the exact order they are given. Do not skip ahead!

STEP 1 Pick a whole number 0–5.

STEP 2 Multiply the number by 3.

STEP 3 Square that product.

STEP 4 Add the digits in your result until you only have 1 digit. For example, 64 → 6 + 4 = 10 → 1 + 0 = 1.

STEP 5 If your sum is less than 5, add 5. If it is greater than 5, subtract 4.

STEP 6 Multiply your new sum or difference by 2.

STEP 7 Subtract 6 from that product.

STEP 8 Assign your new difference a letter in the alphabet starting with 1 = A, 2 = B, 3 = C, and so on.

STEP 9 Pick a state in the United States that begins with your letter.

STEP 10 Now look at the second letter in the name of your chosen state. Choose an animal that begins with that letter.

STEP 11 Share the state and animal you chose with a classmate. How do your choices compare? How do the numbers you chose in Step 1 compare?

Name _____ Date _____ Class _____

LESSON 2-2 Problem Solving
Translate Between Words and Math

Write the correct answer.

1. Holly bought 10 comic books. She gave a few of them to Kyle. Let c represent the number of comic books she gave to Kyle. Write an expression for the number of comic books Holly has left.

   _____

2. Last week, Peter worked 40 hours for $15 an hour. Write a numerical expression for the total amount Peter earned last week. Write an algebraic expression to show how much Peter earns in h hours at that rate.

   _____

3. The temperature dropped 5°F, and then it went up 3°F. Let t represent the beginning temperature. Write an expression to show the ending temperature.

   _____

4. Teri baked 48 cookies and divided them evenly into bags. Let n represent the number of cookies Teri put in each bag. Write an expression for the number of bags she filled.

   _____

Circle the letter of the correct answer.

5. Marisa purchased canned soft drinks for a family reunion. She purchased 1 case of 24 cans and several packages containing 6 cans each. If p represents the number of 6-can packages she purchased, which of the following expressions represents the total number of cans Marisa purchased for the reunion?

   A $24 + 6p$
   B $24 - 6p$
   C $6 + 24p$
   D $6 - 24p$

6. Becky has the addresses of many people listed in her e-mail address book. She forwarded a copy of an article to all but 5 of those people. If a represents the number of addresses, which of the following expressions represents how many people she sent the article to?

   F $a + 5$
   G $5a$
   H $a - 5$
   J $a \div 5$

7. Mei bought several CDs for $12 each. Which of the following expressions could you use to find the total amount she spent on the CDs?

   A $12 + x$
   B $12 - x$
   C $12x$
   D $12 \div x$

8. Tony bought 2 packs of 50 plates and 1 pack of 30 plates. Which of the following expressions could you use to find the total number of plates Tony bought?

   F $2 + 50 + 30$
   G $(2 \cdot 50) + 30$
   H $(2 \cdot 30) + 50$
   J $2(30 + 50)$

Holt Mathematics

Name _____ Date _____ Class _____

LESSON 2-2 Reading Stategies
Use a Visual Map

Identifying word phrases for different operations can help you write algebraic expressions. This visual map shows the four different operations with key word phrases.

$x + 15$	$4 \cdot y$ or $(4)(y)$ or $4y$
• x plus 15 • add 15 to x • the sum of x and 15 • 15 more than x • x increased by 15	• 4 times y • y multiplied by 4 • the product of 4 and y

Word Phrases for Algebraic Expressions

$s - 6$	$\frac{a}{2}$ or $a \div 2$
• 6 subtracted from s • subtract 6 from s • 6 less than s • s decreased by 6 • take away 6 from s	• a divided by 2 • the quotient of a with a divisor of 2

Write a word phrase for each algebraic expression.

1. $t - 8$ _____

2. $\frac{n}{6}$ _____

3. $5w$ _____

4. $z + 12$ _____

Write an algebraic expression for each word phrase.

5. the product of x and 12 _____

6. m decreased by 5 _____

7. the quotient of p with a divisor of 3 _____

8. 25 more than r _____

Name _____ Date _____ Class _____

LESSON 2-2 Puzzles, Twisters & Teasers
Double Time

How many hours did the world's longest doubles table tennis match last?

To find the answer:

1. Use a ruler to match each expression with a phrase for that expression. (Each line you draw will cross a number and a letter.)
2. Write the letter under the matching number in the decoder.

Expression		Phrase
$a + 55$		x divided by 19
$70 \div 35$	13 D	the quotient of 70 and 35
$500 - 125$	7	55 times a
$k + 73$	H E N	the sum of a and 55
$\dfrac{x}{19}$	1 12	
	3 E	313 minus 262
$313 - 262$	10	take away 125 from 500
$x \cdot 19$	N 6	the product of x and 19
$70 + 35$	4 R 5	35 added to 70
$k \cdot 73$	D U	16 times n
$500 \div 125$	2 8	16 plus n
$55a$	E 9	73 more than k
$16n$	N O	k times 73
$16 + n$	11	500 divided by 125

DECODER

7	4	1	12	8	9	3	5	11	13	2	6	10

Copyright © by Holt, Rinehart and Winston.
All rights reserved.

Holt Mathematics

Name _____ Date _____ Class _____

Practice A
LESSON 2-3
Translating Between Tables and Expressions

Circle the letter of the correct answer.

1. Which sentence about the table is true?

Cars	Wheels
1	4
2	8
3	12
c	4c

  A The number of wheels is the number of cars plus 4.
  B The number of wheels is the number of cars minus 4.
  C The number of wheels is the number of cars divided by 4.
  D The number of wheels is 4 times the number of cars.

2. Which sentence about the table is not true?

Brett's Age	Joy's Age
10	11
11	12
12	13
b	b + 1

  F Joy's age is Brett's age plus 1.
  G When Brett's age is b, Joy's age is $b + 1$.
  H Add 1 to Brett's age to get Joy's age.
  J Subtract 1 from Brett's age to get Joy's age.

Write an expression for the missing value in each table.

3.
Motorcycle	Wheels
1	2
2	4
3	6
m	

4.
Marbles	Bags
15	3
20	4
25	5
m	

Write an expression for the sequence in the table.

5.
Position	1	2	3	4	5	n
Value of Term	3	4	5	6	7	

6. What is the value of the term in position 6 in Exercise 5? _____

Name _____ Date _____ Class _____

LESSON 2-3 Practice B
Translating Between Tables and Expressions

Write an expression for the missing value in each table.

1.
Bicycles	Wheels
1	2
2	4
3	6
b	

2.
Ryan's Age	Mia's Age
14	7
16	9
18	11
r	

3.
Minutes	Hours
60	1
120	2
180	3
m	

4.
Bags	Potatoes
3	21
4	28
5	35
b	

Write an expression for the sequence in each table.

5.
Position	1	2	3	4	5	n
Value of Term	3	4	5	6	7	

6.
Position	1	2	3	4	5	n
Value of Term	5	9	13	17	21	

7. A rectangle has a width of 6 inches. The table shows the area of the rectangle for different widths. Write an expression that can be used to find the area of the rectangle when its length is l inches.

Width (in.)	Length (in.)	Area (in.^{2})
6	8	48
6	10	60
6	12	72
6	l	

LESSON 2-3 Practice C
Translating Between Tables and Expressions

Write an expression for the missing value in each table.

1.
Game	Cards
1	52
2	104
3	156
g	

2.
Paper Clips	Boxes
250	5
500	10
750	15
c	

Write an expression for the sequence in each table.

3.
Position	1	2	3	4	5	n
Value of Term	8	13	18	23	28	

4.
Position	1	2	3	4	5	n
Value of Term	0	1	2	3	4	

5. What is the relationship between the number of bags and the number of coins?

_____

_____

Bags	Coins
3	18
5	30
7	42
b	?

6. Look at the table below.

Position	1	2	3	4	5	n
Value of Term	1	3	5	7	9	?

What is the relationship between the positions and the values of the terms in the sequence?

_____

_____

Holt Mathematics

Name _____ Date _____ Class _____

LESSON 2-3 Reteach
Translating Between Tables and Expressions

You can write an expression for data in a table.
The expression must work for all of the data.

Cats	Legs
1	4
2	8
3	12
c	?

Think: When there is 1 cat, there are 4 legs. $4 \times 1 = 4$
When there are 2 cats, there are 8 legs. $4 \times 2 = 8$
When there are 3 cats, there are 12 legs. $4 \times 3 = 12$

So, when there are c cats, there are $4c$ legs.

You can write an expression for the sequence in a table.
Find a rule for the data in the table that works for the whole sequence.

Position	1	2	3	4	5	n
Value of Term	4	5	6	7	8	?

Step 1 Look at the value of the term in position 1.
   4 is **3 more** than 1.
Step 2 Try the rule for position 2.
   5 is **3 more** than 2.
Step 3 Try the rule for the rest of the positions.
   6 is **3 more** than 3, 7 is **3 more** than 4, and 8 is **3 more** than 5.

So, the expression for the sequence is $n + 3$.

Write an expression for the missing value in each table.

1.

People	Legs
1	2
2	4
3	6
p	

2.

Yoko's Age	Mel's Age
9	19
10	20
11	21
y	

Write an expression for the sequence in the table.

3.

Position	1	2	3	4	5	n
Value of Term	3	6	9	12	15	

Name _____ Date _____ Class _____

Challenge

LESSON 2-3 *Table Match*

Match each expression with the table it belongs in.

1.
Games	Players
3	12
4	16
5	20
n	?

2.
Jay's Height (in.)	Lara's Height (in.)
50	47
52	49
54	52
n	?

3.
Photos	Album Pages
27	9
36	12
45	15
n	?

4.
Position	1	2	3	4	n
Value of Term	5	7	9	11	?

5.
Position	1	2	3	4	n
Value of Term	1	3	5	7	?

6.
Position	1	2	3	4	n
Value of Term	2	4	6	8	?

$2n + 3$

$4n$

$4n \div 2$

$n - 3$

$2n - 1$

$n \div 3$

Copyright © by Holt, Rinehart and Winston. All rights reserved.

23

Holt Middle School Math Course 1

LESSON 2-3

Problem Solving
Translating Between Tables and Expressions

Use the table to write an expression for the missing value.
Then use your expression to answer the questions.

1. How many cars are produced on average each year?

   _____

2. How many cars will be produced in 6 years?

   _____

3. After how many years will there be an average production of 3,750 cars?

   _____

Cars Produced By Company X

Number of Years	Average Number of Cars Produced
2	2,500
5	6,250
7	8,750
10	12,500
12	15,000
14	17,500
n	

Circle the letter of the correct answer.
Company Y produces twice as many cars as Company X.

4. How many cars does Company Y produce on average in 8 years?
   A 1,250
   B 10,000
   C 11,250
   D 20,000

5. How many more cars on average does Company Y produce in 4 years than Company X?
   F 2,500
   G 5,000
   H 6,125
   J 7,500

6. Which company produces an average of 11,250 cars in 9 years?
   A Company X
   B Company Y
   C both companies
   D neither company

7. How many cars are produced on average by both companies in 20 years?
   F 3,750
   G 12,500
   H 25,000
   J 37,500

Name _____ Date _____ Class _____

Reading Strategies
2-3 Identify Relationships

When you are **related** to someone, you are connected by something in common. When you look at the positions and the values of terms in a table, they are related, too. You can find the connection, or **relationship**. Then you can write an expression for the sequence.

Position	1	2	3	4	5	n
Value of Term	10	11	12	13	14	?

Read the value of the term in the first position. Note how it is related to its position.

 1   + 9   =   10
 ↑   ↑   ↑
 position 1   relationship between   value of term
             position and value of term   in position 1

Check to see if the relationship works for the second position.

 2   + 9   =   11
 ↑   ↑   ↑
 position 2   relationship between   value of term
             position and value of term   in position 2

Check again. Use the value of the term in the third position.

 3   + 9   =   12
 ↑   ↑   ↑
 position 3   relationship between   value of term
             position and value of term   in position 3

So, the expression for the sequence is $n + 9$.

Use this table to answer Exercises 1–6.

Position	1	2	3	4	5	n
Value of Term	5	10	15	20	25	?

1. How do you go from position 1 to the value of its term, 5? _____

2. Try the relationship for the next term. Can you add 4 to 2 and get 10? _____
   Does $n + 4$ work? _____

3. Try another relationship for postion 1 and its term.
   How else can you go from 1 to 5? _____

4. Try this relationship for the next term. Can you multiply 2 by 5 and get 10? _____

5. Check again by using the value of the term in the third position.
   Can you multiply 3 by 5 and get 15? _____

6. What is the expression for the sequence in the table? _____

Name _____ Date _____ Class _____

LESSON 2-3 Puzzles, Twisters & Teasers
Weather Forecast

What kind of animal can forecast the weather?

Use the table below to answer Exercises 1–8. Circle the correct answer. Then put the letter above the correct answer on the lines.

Position	1	2	3	4	5	n
Value of Term	4	7	10	13	16	?

1. What is the value of the 6th term?
   - K 16
   - A 17
   - B 18
   - R 19

2. What is the value of the 7th term?
   - P 18
   - J 20
   - E 22
   - O 24

3. What is the value of the 8th term?
   - L 20
   - I 25
   - F 30
   - D 35

4. What is the value of the 9th term?
   - X 21
   - C 22
   - N 28
   - H 38

5. What is the value of the 10th term?
   - D 31
   - S 41
   - W 51
   - U 61

6. What is the value of the 11th term?
   - L 24
   - Q 30
   - E 34
   - Z 43

7. What is the value of the 12th term?
   - B 34
   - G 35
   - M 36
   - E 37

8. What is the value of the 13th term?
   - R 40
   - T 42
   - A 44
   - Y 46

___ ___ ___ ___ ___ ___ ___ ___
1.   2.   3.   4.   5.   6.   7.   8.

Name _____ Date _____ Class _____

LESSON 2-4 Practice A
Equations and Their Solutions

Determine whether the given value of the variable is a solution.

1. $x + 1 = 5$ for $x = 4$ _____

2. $13 - w = 10$ for $w = 2$ _____

3. $2 \cdot v = 12$ for $v = 10$ _____

4. $14 \div p = 2$ for $p = 7$ _____

5. $8 + w = 11$ for $w = 3$ _____

6. $4t = 20$ for $t = 5$ _____

7. $\frac{12}{s} = 4$ for $s = 3$ _____

8. $6 + d = 15$ for $d = 8$ _____

Circle the letter of the equation that each given solution makes true.

9. $x = 5$
   A $2 + x = 7$
   B $9 - x = 3$
   C $3 \cdot x = 18$
   D $20 \div x = 2$

10. $g = 7$
    F $9g = 16$
    G $8 - g = 1$
    H $11 + g = 17$
    J $g \div 1 = 1$

11. $y = 2$
    A $5 + y = 8$
    B $7 - y = 1$
    C $3 \cdot y = 6$
    D $10 \div y = 20$

12. $m = 9$
    F $10 + m = 20$
    G $m - 4 = 13$
    H $7 \cdot m = 36$
    J $18 \div m = 2$

13. $z = 4$
    A $5z = 20$
    B $12 \div z = 4$
    C $z - 3 = 7$
    D $z + 8 = 4$

14. $a = 8$
    F $2a = 10$
    G $a + 12 = 20$
    H $a \div 4 = 4$
    J $12 - a = 6$

15. Emanuel put 12 marbles on one pan of a scale. On the other pan, he put 4 marbles, then he added 8 more marbles to that side. Each of the marbles weighs 1 ounce. Is the scale balanced? Explain.

_____

_____

16. Bill and Rhonda have the same amount of money. Bill has $13. Rhonda has one $5 bill, three $1 bills, and one other bill. Is it a $1 bill or a $5 bill? Explain.

_____

_____

Name _____ Date _____ Class _____

LESSON 2-4 Practice B
Equations and Their Solutions

Determine whether the given value of the variable is a solution.

1. $9 + x = 21$ for $x = 11$ _____

2. $n - 12 = 5$ for $n = 17$ _____

3. $25 \cdot r = 75$ for $r = 3$ _____

4. $72 \div q = 8$ for $q = 9$ _____

5. $28 + c = 43$ for $c = 15$ _____

6. $u \div 11 = 10$ for $u = 111$ _____

7. $\dfrac{k}{8} = 4$ for $k = 24$ _____

8. $16x = 48$ for $x = 3$ _____

9. $73 - f = 29$ for $f = 54$ _____

10. $67 - j = 25$ for $j = 42$ _____

11. $39 \div v = 13$ for $v = 3$ _____

12. $88 + d = 100$ for $d = 2$ _____

13. $14p = 20$ for $p = 5$ _____

14. $6w = 30$ for $w = 5$ _____

15. $7 + x = 70$ for $x = 10$ _____

16. $6 \cdot n = 174$ for $n = 29$ _____

Replace each ? with a number that makes the equation correct.

17. $5 + 1 = 2 + \boxed{?}$ _____

18. $10 - \boxed{?} = 12 - 7$ _____

19. $\boxed{?} \cdot 3 = 2 \cdot 9$ _____

20. $28 \div 4 = 14 \div \boxed{?}$ _____

21. $\boxed{?} + 8 = 6 + 3$ _____

22. $12 \cdot 0 = \boxed{?} \cdot 15$ _____

23. Carla had $15. After she bought lunch, she had $8 left. Write an equation using the variable x to model this situation. What does your variable represent?

24. Seventy-two people signed up for the soccer league. After the players were evenly divided into teams, there were 6 teams in the league. Write an equation to model this situation using the variable x.

Name _____ Date _____ Class _____

LESSON 2-4 Practice C
Equations and Their Solutions

For each equation, determine whether the given value of the variable is a solution.

1. $2d = 24$ for $d = 8$ _____
2. $15 \div p = 6$ for $p = 3$ _____
3. $x^2 = 25$ for $x = 5$ _____
4. $135 \div x = 9$ for $x = 15$ _____
5. $7 + t - 4 = 22$ for $t = 20$ _____
6. $\frac{4s}{8} = 6$ for $s = 12$ _____
7. $\frac{u}{13} = 5$ for $u = 65$ _____
8. $2x + 3x = 60$ for $x = 12$ _____
9. $2(100 - k) = 64$ for $k = 36$ _____
10. $(69 \div m) - 5 = 18$ for $m = 3$ _____
11. $11 \div w = 11$ for $w = 1$ _____
12. $576 + n = 1{,}000$ for $n = 524$ _____
13. $(15 \cdot 6)y = 450$ for $y = 5$ _____
14. $18c \div 2 = 89$ for $c = 11$ _____
15. $6^2 - r = 20$ for $r = 26$ _____
16. $x^5 = 32$ for $x = 2$ _____

Replace each ? with a number that makes the equation correct.

17. $28 + 11 = 22 + ?$ ____
18. $19 - ? = 75 - 60$ ____
19. $? \cdot 12 = 21 \cdot 4$ ____
20. $54 \div 6 = 108 \div ?$ ____
21. $? + 87 = 46 + 59$ ____
22. $3^2 \cdot 3 = ? \cdot 3^3$ ____

23. Mr. Yakima teaches 4 science classes, with the same number of students in each class. Of those students, 80 are sixth graders, and 40 are fifth graders. Write an equation to model this situation using the variable n. What does your variable represent?

_____

_____

24. Mary put new tiles on her kitchen floor. The floor measures 6 feet long by 5 feet wide. She used 10 tiles to cover the entire floor. Each tile was 3 feet long. Write an equation using the variable x to model this situation. What does x represent in the equation?

_____

_____

Name _____ Date _____ Class _____

LESSON 2-4 Reteach
Equations and Their Solutions

An equation is a mathematical sentence that says that two quantities are equal.

Some equations contain variables. A solution for an equation is a value for a variable that makes the statement true.

You can write related facts using addition and subtraction.
$7 + 6 = 13$ $13 - 6 = 7$

You can write related facts using multiplication and division.
$3 \cdot 4 = 12$ $12 \div 4 = 3$

You can use related facts to find solutions for equations. If the related fact matches the value for the variable, then that value is a solution.

A. $x + 5 = 9$, when $x = 3$
    Think: $9 - 5 = x$
         $x = 4$
         $3 \neq 4$
    So $x = 3$ is not a solution of $x + 5 = 9$.

B. $x - 7 = 5$, when $x = 12$
    Think: $5 + 7 = x$
         $x = 12$
         $12 = 12$
    So $x = 12$ is a solution of $x - 7 = 5$.

C. $2x = 14$, when $x = 9$
    Think: $14 \div 2 = x$
         $x = 7$
         $9 \neq 7$
    So $x = 9$ is not a solution for $2x = 14$.

D. $x \div 5 = 3$, when $x = 15$
    Think: $3 \cdot 5 = x$
         $x = 15$
         $15 = 15$
    So $x = 15$ is a solution for $x \div 5 = 3$.

Use related facts to determine whether the given value is a solution for each equation.

1. $x + 6 = 14$, when $x = 8$ 2. $s \div 4 = 5$, when $s = 24$ 3. $g - 3 = 7$, when $g = 11$

_____   _____   _____

4. $3a = 18$, when $a = 6$ 5. $26 = y - 9$, when $y = 35$ 6. $b \cdot 5 = 20$, when $b = 3$

_____   _____   _____

7. $15 = v \div 3$, when $v = 45$ 8. $11 = p + 6$, when $p = 5$ 9. $6k = 78$, when $k = 12$

_____   _____   _____

Copyright © by Holt, Rinehart and Winston.
All rights reserved.

Holt Mathematics

Name _____ Date _____ Class _____

LESSON 2-4 Challenge
Keep It Balanced

Study the scales below. Then circle the solution below each scale that will keep it balanced.

1.
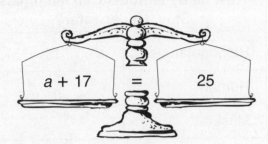

$a = 8$ $a = 9$

2.
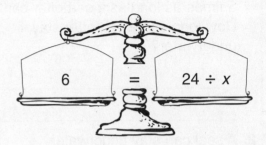

$x = 3$ $x = 4$

3.
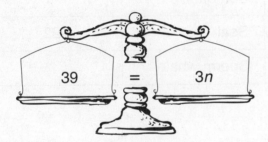

$n = 12$ $n = 13$

4.
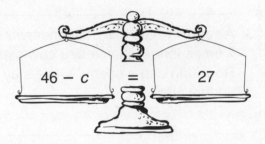

$c = 19$ $c = 29$

5.
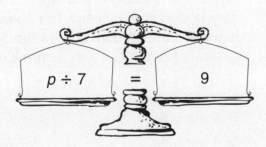

$p = 49$ $p = 63$

6.
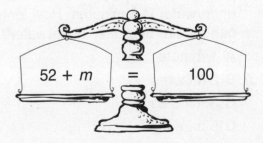

$m = 48$ $m = 58$

Copyright © by Holt, Rinehart and Winston.
All rights reserved.

Holt Mathematics

Problem Solving
LESSON 2-4 Equations and Their Solutions

Use the table to write and solve an equation to answer each question. Then use your answers to complete the table.

1. A hippopotamus can stay underwater 3 times as long as a sea otter can. How long can a sea otter stay underwater?

   _____

2. A seal can stay underwater 10 minutes longer than a muskrat can. How long can a muskrat stay underwater?

   _____

3. A sperm whale can stay underwater 7 times longer than a sea cow can. How long can a sperm whale stay underwater?

   _____

How Many Minutes Can Mammals Stay Underwater?	
Hippopotamus	15
Human	
Muskrat	
Platypus	10
Polar bear	
Sea cow	16
Sea otter	
Seal	22
Sperm whale	

Circle the letter of the correct answer.

4. The difference between the time a platypus and a polar bear can stay underwater is 8 minutes. How long can a polar bear stay underwater?
   A 1 minute
   B 2 minutes
   C 3 minutes
   D 5 minutes

5. When you divide the amount of time any of the animals in the table can stay underwater by itself, the answer is always the amount of time the average human can stay underwater. How long can the average human stay underwater?
   F 6 minutes
   G 4 minutes
   H 2 minutes
   J 1 minute

Name _____ Date _____ Class _____

LESSON 2-4 Reading Strategies
Focus on Vocabulary

You can see the word **equal** in **equation**. In math, an equation indicates that two quantities are equal, or the same. The = **sign** in an equation separates one quantity from the other. The value on each side of the = sign is the same.

Look at the equations below. Notice how the value on each side of the = sign is the same for each equation:

$5 + 7 = 8 + 4$ $19 - 7 = 12$ $42 = 3 \cdot 14$

If an equation contains a variable, and the variable is replaced by a value that keeps the equation equal, that value is called a **solution** of the equation.

Examples:
$y \div 4 = 15$ $y \div 4 = 15$
$80 \div 4 \neq 15$ $60 \div 4 = 15$
"80 divided by 4 is not equal to 15."   "60 divided by 4 is equal to 15."

Which are equations? Write the correct sign, = or ≠.

1. $7 + 23$ ☐ $9 + 21$ _____

2. $35 + 15$ ☐ 45 _____

3. $28 - 7$ ☐ $15 + 6$ _____

Replace the given value for the variable. Is it a solution?

4. $d + 28 = 45$ for $d = 17$ _____

5. $84 \div s = 28$ for $s = 3$ _____

6. $17 = 56 - t$ for $t = 40$ _____

7. $86 = 4w$ for $w = 24$ _____

Puzzles, Twisters & Teasers
Lesson 2-4 Space Fact

On July 4, 1997, what did the Pathfinder spacecraft do?

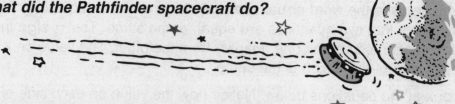

For each equation, determine whether the given value of the variable is a solution. If it is a solution, circle =. If it is not a solution, circle ≠. Put the letter above the correct answer in the box.

#	Equation	=	≠	Box
1.	$117 = 97 + n$ for $n = 10$	A	**I**	I
2.	$96 \div x = 8$ for $x = 12$	**T**	S	T
3.	$132 \div m = 12$ for $m = 12$	K	**L**	L
4.	$k + 18 = 63$ for $k = 44$	E	**A**	A
5.	$35 \div s = 7$ for $s = 5$	**N**	M	N
6.	$44 = t - 55$ for $t = 88$	J	**D**	D
7.	$a - 6 = 36$ for $a = 42$	**E**	A	E
8.	$u \cdot 7 = 72$ for $u = 8$	P	**D**	D
9.	$b + 21 = 28$ for $b = 8$	I	**O**	O
10.	$92 - 28 = 8y$ for $y = 8$	**N**	F	N
11.	$6x = 54$ for $x = 9$	**M**	P	M
12.	$149 = 79 + 2y$ for $y = 35$	**A**	E	A
13.	$17w - 50 = 0$ for $w = 3$	T	**R**	R
14.	$25g = 25{,}000$ for $g = 1{,}000$	**S**	H	S

Answer: IT LANDED ON MARS

Name _____ Date _____ Class _____

LESSON 2-5
Practice A
Addition Equations

Match each equation in Column 1 to its solution in Column 2.

Column 1

1. $5 + x = 8$ _____
2. $12 + x = 12$ _____
3. $x + 11 = 15$ _____
4. $x + 9 = 20$ _____
5. $8 + x = 13$ _____
6. $6 + x = 14$ _____
7. $2 + x = 11$ _____
8. $x + 29 = 30$ _____
9. $3 + x = 10$ _____
10. $x + 17 = 19$ _____

Column 2

A. $x = 5$
B. $x = 3$
C. $x = 11$
D. $x = 9$
E. $x = 7$
F. $x = 2$
G. $x = 4$
H. $x = 8$
I. $x = 0$
J. $x = 1$

Solve each equation. Check your answers.

11. $p + 8 = 14$

12. $q + 10 = 13$

13. $7 + s = 15$

14. $4 + w = 11$

15. $t + 12 = 15$

16. $9 + m = 14$

17. Phyllis has 6 yards of material. She needs 8 yards to make curtains. This situation is modeled by the equation $6 + x = 8$, where x is the amount of material she needs to buy. How much more material does she need to buy to make the curtains?

18. Emma paid $26 in all for a hammer and a screwdriver. The hammer cost $10. Write an addition equation using the variable n to show how much she spent on the screwdriver.

LESSON 2-5 Practice B
Addition Equations

Solve each equation. Check your answers.

1. $s + 3 = 23$

2. $v + 10 = 49$

3. $q + 9 = 16$

4. $81 + m = 90$

5. $38 + x = 44$

6. $28 + n = 65$

7. $t + 31 = 50$

8. $25 + p = 39$

9. $19 + v = 24$

Solve each equation.

10. $m + 8 = 17$

11. $r + 14 = 20$

12. $25 + x = 32$

13. $47 + p = 55$

14. $19 + d = 27$

15. $13 + n = 26$

16. $q + 12 = 19$

17. $34 + f = 43$

18. $52 + w = 68$

19. Kenya bought 28 beads, and Nancy bought 25 beads. It takes 35 beads to make a necklace. Write and solve two addition equations to find how many more beads they each need to make a necklace.

20. During a sales trip, Mr. Jones drove 15 miles east from Brownsville to Carlton. Then he drove several more miles east from Carlton to Sun City. The distance from Brownsville to Sun City is 35 miles. Write and solve an addition equation to find how many miles it is from Carlton to Sun City.

Name _____ Date _____ Class _____

LESSON 2-5
Practice C
Addition Equations

Solve each equation. Check your answers.

1. $s + 67 = 101$

2. $v + 13 = 28 - 5$

3. $29 + q + 18 = 51$

4. $4^2 + m = 35$

5. $78 + x = 121 - 4$

6. $6 + n = 28 - 9$

7. $t + 1{,}906 = 2{,}000$

8. $41 + p + 16 = 99$

9. $201 + v + 30 = 249$

Solve each equation.

10. $m + 38 = 90$

11. $12 + r + 17 = 60$

12. $115 + x = 320$

13. $57 + p = 63 + 18$

14. $2^3 + d = 21$

15. $15 + n = 8^2$

16. $q + 6 + 77 = 100$

17. $13{,}687 + t = 20{,}441$

18. $25 + w + 2 = 37$

19. Alice bought a round-trip ticket to fly from Baltimore to Chicago on SuperAir for $250. That was $16 more than she would have paid on Jet Airlines, which only offered a one-way fair. How much did Jet Airlines charge to fly from Baltimore to Chicago?

20. Together, Ken, Judy, Sam, and Ali worked a total of 83 hours last week. Ken worked 3 more hours than Judy. Ali worked 5 more hours than Ken. Judy worked 4 more hours than Sam. If Sam worked 15 hours, how many hours did Ken, Judy, and Ali each work?

LESSON 2-5 Reteach
Addition Equations

To solve an equation, you need to get the variable alone on one side of the equal sign.

You can use tiles to help you solve addition equations.

Subtraction undoes addition, so you can use subtraction to solve addition equations.

variable   add 1   subtract 1   add 1   subtract 1   ← zero pair

One positive tile and one negative tile is called a zero pair because together they have a value of zero.

To solve $x + 3 = 5$, first use tiles to model the equation.

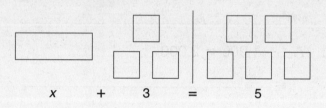

Next, add enough subtraction tiles to get the variable alone. Then add the same number of subtraction tiles to the other side of the equal sign.

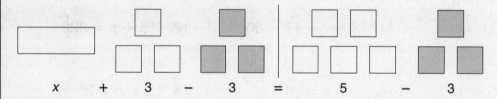

Then remove the greatest possible number of zero pairs from each side of the equal sign.

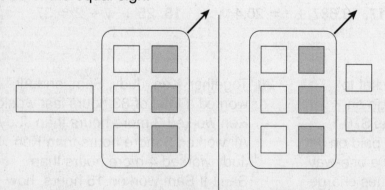

Check: $x + 3 = 5$
$2 + 3 \stackrel{?}{=} 5$
$5 \stackrel{?}{=} 5$ ✓

The remaining tiles represent the solution.
$x = 2$

Use tiles to solve each equation. Then check each answer.

1. $x + 8 = 11$
2. $x + 4 = 9$
3. $x + 7 = 13$

Name _____ Date _____ Class _____

Challenge
LESSON 2-5   The Temperature's Rising

Each pair of thermometers shows a beginning temperature on the left and an ending temperature on the right. Write and solve an addition equation to find the change in temperature shown on each pair of thermometers.

1.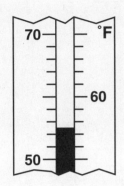

_____

2.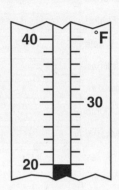

_____

3.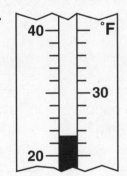

_____

4.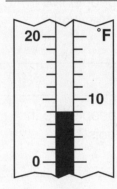

_____

5.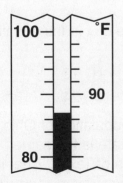

_____

6.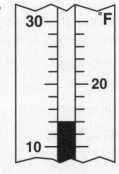

_____

Holt Mathematics

Name _____ Date _____ Class _____

LESSON 2-5 Problem Solving
Addition Equations

Use the bar graph and addition equations to answer the questions.

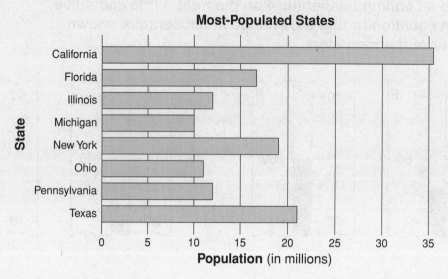

1. How many more people live in California than in New York?

   _____

2. How many more people live in Ohio than in Michigan?

   _____

3. How many more people live in Florida than in Illinois?

   _____

4. How many more people live in Texas than in Pennsylvania?

   _____

Circle the letter of the correct answer.

5. Which two states' populations are used in the equation $12 + x = 12$?

   A Pennsylvania and Texas
   B Ohio and Florida
   C Michigan and Illinois
   D Illinois and Pennsylvania

6. What is the value of x in the equation in Exercise 5?

   F 0   H 12
   G 1   J 24

7. In 2003, the total population of the United States was 292 million. How many of those people did not live in one of the states shown on the graph?

   A 416 million   C 154 million
   B 73 million   D 292 million

8. The combined population of Ohio and one other state is the same as the population of Texas. What is that state?

   F California
   G Florida
   H Michigan
   J Pennsylvania

Copyright © by Holt, Rinehart and Winston.
All rights reserved.

Holt Mathematics

LESSON 2-5

Reading Strategies
Use a Visual Cue

An equation is like a balance scale. The value on the right side of the balance scale or equation is equal to the value on the left side of the balance scale or equation.

A balanced scale also helps you to picture a balanced equation:

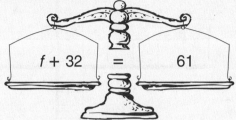

Step 1: To find the value of f, the variable needs to be by itself on one side of the equation. So 32 must be subtracted from the left side of the equation.

Step 2: To keep the scale balanced, subtract 32 from the right side of the equation as well.

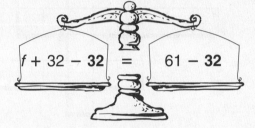

Step 3: Check to verify that $f = 29$ is the solution.

$f + 32 = 61$

$29 + 32 \stackrel{?}{=} 61$

$61 \stackrel{?}{=} 61$ ✓   29 is the solution.

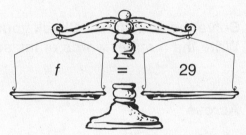

To get the variable by itself in an addition equation, subtract the same value from both sides of the equation.

Use $z + 16 = 42$ to answer Exercises 1–4.

1. On which side of the equation is the variable? _____

2. What will you do to get the variable by itself? _____

3. What must you do to the other side of the equation to keep it balanced? _____

4. What is the value of z? _____

Name _____ Date _____ Class _____

LESSON 2-5 Puzzles, Twisters & Teasers
Numbers "r" Us

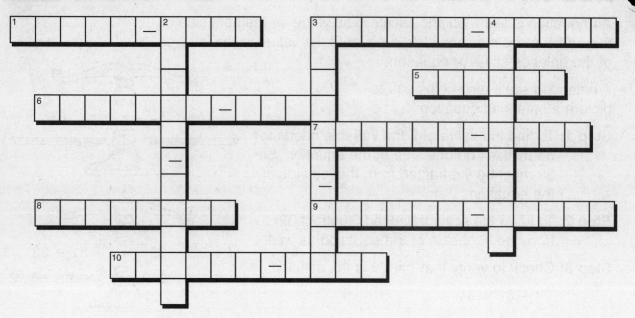

Solve the equations. Check your work.
Write the number words of the solution in the puzzle.

Across

1. $31 + n = 75$
3. $75 = m + 38$
5. $29 = 18 + c$
6. $77 + t = 151$
7. $w + 52 = 81$
8. $57 + z = 71$
9. $a + 32 = 60$
10. $40 = p + 16$

Down

2. $b + 96 = 153$
3. $x + 15 = 47$
4. $z + 89 = 106$

Copyright © by Holt, Rinehart and Winston.
All rights reserved.

42

Holt Mathematics

Practice A
LESSON 2-6 Subtraction Equations

Match each equation in Column 1 to its solution in Column 2.

Column 1

1. $x - 3 = 1$ _____
2. $3 = x - 5$ _____
3. $x - 1 = 6$ _____
4. $2 = x - 4$ _____
5. $x - 2 = 3$ _____
6. $4 = x - 6$ _____
7. $x - 8 = 1$ _____
8. $7 = x - 5$ _____
9. $x - 1 = 2$ _____
10. $9 = x - 2$ _____

Column 2

A. $x = 5$
B. $x = 3$
C. $x = 10$
D. $x = 9$
E. $x = 7$
F. $x = 11$
G. $x = 4$
H. $x = 8$
I. $x = 12$
J. $x = 6$

Solve each equation. Check your answers.

11. $p - 6 = 4$

12. $q - 3 = 9$

13. $19 = s - 2$

14. $7 = w - 4$

15. $5 = t - 8$

16. $m - 11 = 6$

17. A box of pencils costs $3. Brian got $2 change after paying for one box. Write a subtraction equation using the variable x to show how much Brian gave the cashier.

18. After dropping 5°F, the temperature was 25°F. Write a subtraction equation using the variable t to show what the starting temperature was.

Name _____ Date _____ Class _____

LESSON 2-6
Practice B
Subtraction Equations

Solve each equation. Check your answers.

1. $s - 8 = 12$

2. $v - 11 = 7$

3. $9 = q - 5$

4. $m - 21 = 5$

5. $34 = x - 12$

6. $n - 45 = 45$

7. $t - 19 = 9$

8. $p - 6 = 27$

9. $15 = v - 68$

Solve each equation.

10. $7 = m - 5$

11. $r - 10 = 22$

12. $16 = x - 4$

13. $40 = p - 11$

14. $28 = d - 6$

15. $n - 9 = 42$

16. $q - 85 = 8$

17. $f - 13 = 18$

18. $47 = w - 38$

19. Ted took 17 pictures at the aquarium. He now has 7 pictures left on the roll. Write and solve a subtraction equation to find out how many photos Ted had when he went to the aquarium.

20. Ted bought a dolphin poster for $12. He now has $5. Write and solve a subtraction equation to find out how much money Ted took to the aquarium.

Name _____ Date _____ Class _____

LESSON 2-6 Practice C
Subtraction Equations

Solve each equation. Check your answers.

1. $s - 57 = 38$

2. $v - 16 = 12 + 6$

3. $q - 18 - 5 = 20$

4. $m - 3^2 = 15$

5. $159 = x - 78$

6. $n - 4^2 = 4$

7. $t - 4{,}360 = 1{,}804$

8. $p - 63 - 14 = 99$

9. $v - 50 = 14 + 9$

Solve each equation.

10. $m - 79 = 12$

11. $r - 109 = 65$

12. $x - 58 = 370$

13. $p - 16 = 7 + 6$

14. $d - 2^4 = 20$

15. $7^2 = n - 11$

16. $q - 12 - 140 = 15$

17. $t - 18{,}620 = 19{,}000$

18. $w - 3^2 = 16 + 2$

19. If $x - y = 6$, and $x = y + y$, what are the values for x and y?

20. If $x + 4 = y - 2$, and $y = x + x + x$, what are the values for x and y?

Copyright © by Holt, Rinehart and Winston.
All rights reserved.

Holt Mathematics

Name _____ Date _____ Class _____

LESSON 2-6 Reteach
Subtraction Equations

To solve an equation, you need to get the variable alone on one side of the equal sign.

You can use tiles to help you solve subtraction equations.

Addition undoes subtraction, so you can use addition to solve subtraction equations.

   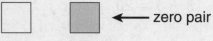
variable   add 1   subtract 1   add 1   subtract 1 ← zero pair

One positive tile and one negative tile is called a zero pair because together they have a value of zero.

To solve $x - 4 = 2$, first use tiles to model the equation.

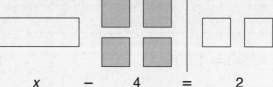

$x \quad - \quad 4 \quad = \quad 2$

Next, add enough addition tiles to get the variable alone. Then add the same number of addition tiles to the other side of the equal sign.

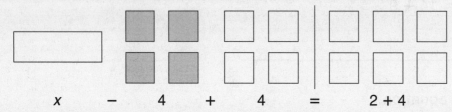

$x \quad - \quad 4 \quad + \quad 4 \quad = \quad 2 + 4$

Then remove the greatest possible number of zero pairs from each side of the equal sign.

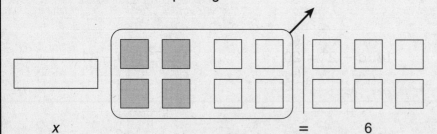

$x \quad\quad = \quad 6$

Check: $x - 4 = 2$
$6 - 4 \stackrel{?}{=} 2$
$2 \stackrel{?}{=} 2$ ✓

The remaining tiles represent the solution.
$x = 6$

Use tiles to solve each equation. Then check each answer.

1. $x - 5 = 3$ _____ 2. $x - 2 = 5$ _____ 3. $x - 6 = 4$ _____

4. $x - 8 = 1$ _____ 5. $x - 3 = 9$ _____ 6. $x - 7 = 3$ _____

Name _____ Date _____ Class _____

Challenge
LESSON 2-6 *The Price Is Right*

Each of the grocery items on this page has a different price—
$1, $2, $3, $4, or $5. Use logic and the subtraction equations
below to figure out the price of each item. Then write the
correct price on each item's price tag.

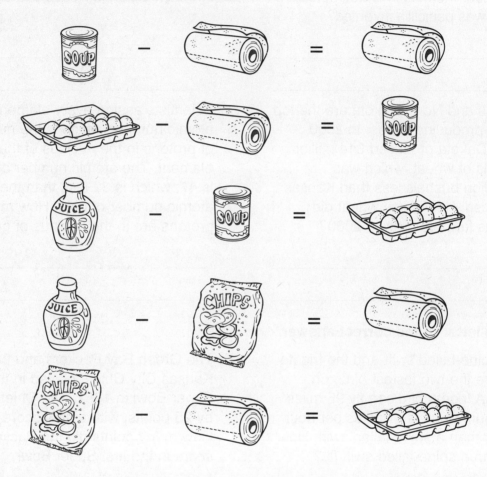

47 Holt Mathematics

Name _____ Date _____ Class _____

LESSON 2-6 Problem Solving
Subtraction Equations

Write and solve subtraction equations to answer the questions.

1. Dr. Felix Hoffman invented aspirin in 1899. That was 29 years before Alexander Fleming invented penicillin. When was penicillin invented?

   _____

2. Kimberly was born on February 2. That is 10 days earlier than Kent's birthday. When is Kent's birthday?

   _____

3. Kansas and North Dakota are the top wheat-producing states. In 2000, North Dakota produced 314 million bushels of wheat, which was 34 million bushels less than Kansas produced. How much wheat did Kansas farmers grow in 2000?

   _____

4. Scientists assign every element an atomic number, which is the number of protons in the nucleus of that element. The atomic number of silver is 47, which is 32 less than the atomic number of gold. How many protons are in the nucleus of gold?

   _____

Circle the letter of the correct answer.

5. The spine-tailed swift and the frigate bird are the two fastest birds on earth. A frigate bird can fly 95 miles per hour, which is 11 miles per hour slower than a spine-tailed swift. How fast can a spine-tailed swift fly?

   A 84 miles per hour
   B 101 miles per hour
   C 106 miles per hour
   D 116 miles per hour

6. The Green Bay Packers and the Kansas City Chiefs played in the first Super Bowl in 1967. The Chiefs lost by 25 points, with a final score of 10. How many points did the Packers score in the first Super Bowl?

   F 35
   G 25
   H 15
   J 0

7. The Rocky Mountains extend 3,750 miles across North America. That is 750 miles shorter than the Andes Mountains in South America. How long are the Andes Mountains?

   A 3,000 miles   C 180 miles
   B 5 miles   D 4,500 miles

8. When the United States took its first census in 1790, only 4 million people lived here. That was 288 million fewer people than the population in 2003. What was the population of the United States in 2003?

   F 292 million   H 69 million
   G 284 million   J 1,108 million

Name _____ Date _____ Class _____

Reading Strategies
2-6 Use a Visual Cue

You can picture balanced scales and follow similar steps to solve subtraction equations.

Picture balanced scales for this equation.

Step 1: To find the value of b, get b by itself on the left side of the equation. So add 17 to the left side of the equation.

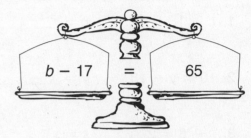

Step 2: To keep the equation balanced, add 17 to the right side of the equation as well.

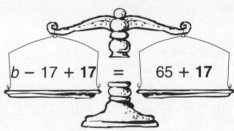

Step 3: Check to verify that $b = 82$ is the solution.
$b - 17 = 65$
$82 - 17 \stackrel{?}{=} 65$
$65 \stackrel{?}{=} 65$ ✓   82 is the solution.

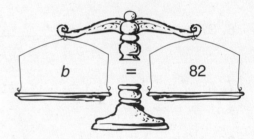

To get the variable by itself in a subtraction equation, add the same value to both sides of the equation.

Use $t - 18 = 53$ to answer Exercises 1–4.

1. On which side of the equation is the variable? _____

2. What will you do to get the variable by itself? _____

3. What must you do to the other side of the equation to keep it balanced? _____

4. What is the value of t? _____

Puzzles, Twisters & Teasers
LESSON 2-6 — Opposites Attract

Why are addition and subtraction like poems?

Solve the following equations to decode the answer.

1. $n - 16 = 12$ $n =$ _____

2. $10 = y - 13$ $y =$ _____

3. $s - 6 = 2$ $s =$ _____

4. $26 = i - 14$ $i =$ _____

5. $g - 82 = 93$ $g =$ _____

6. $37 = a - 5$ $a =$ _____

7. $r - 7 = 23$ $r =$ _____

8. $64 = v - 27$ $v =$ _____

9. $46 = b - 31$ $b =$ _____

10. $e - 5 = 4$ $e =$ _____

They are __ __ __ __ __ __ __ __ .
          40 28 91 9 30 8 9 8

Name _____ Date _____ Class _____

LESSON 2-7 Practice A
Multiplication Equations

Match each equation in Column 1 to its solution in Column 2.

Column 1	Column 2
1. $5x = 40$ _____	A. $x = 7$
2. $21 = 3x$ _____	B. $x = 5$
3. $6x = 24$ _____	C. $x = 3$
4. $42 = 7x$ _____	D. $x = 9$
5. $2x = 18$ _____	E. $x = 4$
6. $20 = 4x$ _____	F. $x = 8$
7. $17x = 17$ _____	G. $x = 0$
8. $16 = 8x$ _____	H. $x = 2$
9. $20x = 0$ _____	I. $x = 1$
10. $27 = 9x$ _____	J. $x = 6$

Solve each equation. Check your answers.

11. $6p = 30$

12. $4q = 12$

13. $21 = 3s$

14. $8 = 2w$

15. $25 = 5t$

16. $9m = 54$

17. Cheryl gets paid $8 per hour at her job at the record store. She made a total of $96 last week. Write a multiplication equation using the variable h to show how many hours she worked last week.

18. There are 3 feet in a yard. John used 27 feet of wire in his sculpture. Write a multiplication equation using the variable y to find how many yards of wire John used in his sculpture.

Name _____ Date _____ Class _____

LESSON 2-7

Practice B
Multiplication Equations

Solve each equation. Check your answers.

1. $8s = 72$

2. $4v = 28$

3. $27 = 9q$

4. $12m = 60$

5. $48 = 6x$

6. $7n = 63$

7. $10t = 130$

8. $15p = 450$

9. $84 = 6v$

Solve each equation.

10. $49 = 7m$

11. $20r = 80$

12. $64 = 8x$

13. $36 = 4p$

14. $147 = 7d$

15. $11n = 110$

16. $12q = 144$

17. $25f = 125$

18. $128 = 16w$

19. A hot-air balloon flew at 10 miles per hour. Using the variable h, write and solve a multiplication equation to find how many hours the balloon traveled if it covered a distance of 70 miles.

20. A passenger helicopter can travel 300 miles in the same time it takes a hot-air balloon to travel 20 miles. Using the variable s, write and solve a multiplication equation to find how many times faster the helicopter can travel than the hot air balloon.

Name _____ Date _____ Class _____

LESSON 2-7 Practice C
Multiplication Equations

Solve each equation. Check your answers.

1. $24s = 144$

2. $5v = \dfrac{225}{15}$

3. $4q = \dfrac{16}{2}$

4. $(3^2)m = 45$

5. $266 = 38x$

6. $(4^2)n = 48$

7. $213t = 1{,}917$

8. $(15p) \cdot 4 = 660$

9. $3v = 300$

Solve each equation.

10. $65m = 845$

11. $105r = 840$

12. $(2^4)x = 112$

13. $(10p) \cdot 21 = 1{,}890$

14. $42d = 210$

15. $36 = 3n \cdot 4$

16. $57 \cdot (4q) = 228$

17. $137t = 822$

18. $(3^3)w = 54$

19. If $xy = 56$, and $y - x = 1$, what are the values for x and y?

20. If $x = 4y$, and $x + y = 10$, what are the values for x and y?

Copyright © by Holt, Rinehart and Winston.
All rights reserved.

Holt Mathematics

LESSON 2-7 Reteach

Multiplication Equations

You can use tiles to help you solve multiplication equations.

Division undoes multiplication, so you can use division to solve multiplication equations.

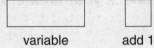
variable   add 1

To solve $3x = 12$, first use tiles to model the equation.

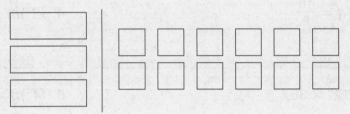

$3x = 12$

Next, divide each side of the equal sign into 3 equal groups

The number of tiles in one group represents the solution.

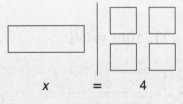

$x = 4$

Check: $3x = 12$
$3 \cdot 4 \stackrel{?}{=} 12$
$12 \stackrel{?}{=} 12$ ✓

$x = 4$

Use tiles to solve each equation. Then check each answer.

1. $5x = 15$ 2. $2x = 6$ 3. $4x = 16$ 4. $8x = 24$

   _____   _____   _____   _____

5. $3x = 18$ 6. $6x = 12$ 7. $7x = 21$ 8. $9x = 9$

   _____   _____   _____   _____

9. $4x = 24$ 10. $3x = 9$ 11. $8x = 16$ 12. $5x = 25$

   _____   _____   _____   _____

Name _____ Date _____ Class _____

LESSON 2-7 Challenge
Regulation Sizes

Write a multiplication equation for the area of each regulation field or court. Then solve the equations to find the missing measurements. Remember: Area = length • width, or $A = l \cdot w$.

1.
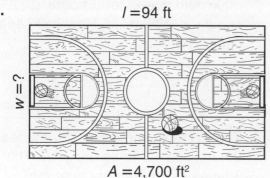
$l = 94$ ft
$w = ?$
$A = 4{,}700$ ft²

What is the width of a regulation basketball court?

_____

2.
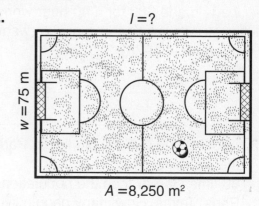
$l = ?$
$w = 75$ m
$A = 8{,}250$ m²

What is the length of a regulation soccer field?

_____

3.
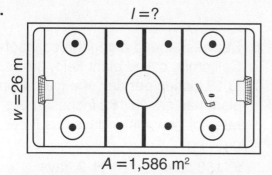
$l = ?$
$w = 26$ m
$A = 1{,}586$ m²

What is the length of a regulation ice hockey rink?

_____

4.
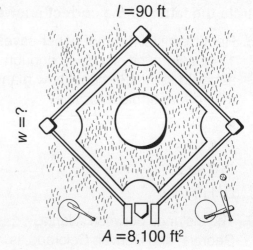
$l = 90$ ft
$w = ?$
$A = 8{,}100$ ft²

What is the width of a regulation baseball diamond?

_____

Copyright © by Holt, Rinehart and Winston.
All rights reserved.

Holt Mathematics

Problem Solving
2-7 Multiplication Equations

Write and solve a multiplication equation to answer each question.

1. In 1975, a person earning minimum wage made $80 for a 40-hour work week. What was the minimum wage per hour in 1975?

2. If an ostrich could maintain its maximum speed for 5 hours, it could run 225 miles. How fast can an ostrich run?

3. About 2,000,000 people live in Paris, the capital of France. That is 80 times larger than the population of Paris, Texas. How many people live in Paris, Texas?

4. The average person in China goes to the movies 12 times per year. That is 3 times more than the average American goes to the movies. How many times per year does the average American go to the movies?

Circle the letter of the correct answer.

5. Recycling just 1 ton of paper saves 17 trees! If a city recycled enough paper to save 136 trees, how many tons of paper did it recycle?
   A 7 tons
   B 8 tons
   C 9 tons
   D 119 tons

6. Seaweed found along the coast of California, called giant kelp, grows up to 18 inches per day. If a giant kelp plant has grown 162 inches at this rate, for how many days has it been growing?
   F 180 days   H 9 days
   G 144 days   J 8 days

7. The distance between Atlanta, Georgia, and Denver, Colorado, is 1,398 miles. That is twice the distance between Atlanta and Detroit, Michigan. How many miles would you have to drive to get from Atlanta to Detroit?
   A 2,796 miles
   B 349.5 miles
   C 699 miles
   D 1,400 miles

8. Jupiter has 2 times more moons than Neptune has, and 8 times more moons than Mars has. Jupiter has 16 moons. How many moons do Neptune and Mars each have?
   F 8 moons, 2 moons
   G 2 moons, 8 moons
   H 128 moons, 32 moons
   J 32 moons, 128 moons

Name _____ Date _____ Class _____

Reading Strategies
LESSON 2-7
Follow a Procedure

Multiplication and division are **inverse operations.** You can think of them as **opposite operations.**

$4 \cdot 12 = 48$ and $48 \div 12 = 4$

$6 \cdot 13 = 78$ and $78 \div 13 = 6$

From these examples, you could say that division **"undoes"** the multiplication.

Follow these steps to "undo" the multiplication and solve.

$7n = 84$ ⟶ Read: "7 times n equals 84."

Step 1: Get n by itself. Use division to "undo" multiplication. Since 7 is multiplied by n, divide by 7.

$7n = 84$

Step 2: To keep the equation balanced, divide the right side of the equation by 7 also.

$7n \div 7 = 84 \div 7$

Step 3: Check to verify that $n = 12$ is the solution.

$n = 12$
$7n = 84$
$7 \cdot 12 \stackrel{?}{=} 84$
$84 \stackrel{?}{=} 84$ ✔ 12 is the solution.

Answer each question.

1. What is another name for the "opposite operation"? _____

2. What is the inverse operation for multiplication? _____

Use $8z = 96$ for Exercises 3–6.

3. Write the equation in words. _____

4. What operation is used in the equation? _____

5. What operation will you perform on both sides of the equation to solve it? _____

6. Solve the equation. _____

Name _____ Date _____ Class _____

LESSON 2-7 Puzzles, Twisters & Teasers
Go for the Gold!

Where did the 2002 Winter Olympics take place?
Find the letter that corresponds to the answer for each exercise using the decoder below. Then, place the letter in the blank corresponding to the exercise number to determine where the 2002 Winter Olympics were held.

1. At the 2002 Winter Olympics, Norway won 24 medals. Norway won 3 times as many medals as the Netherlands. How many medals did the Netherlands win?

   Solve: Check:

2. At the 2002 Games, the United States won 34 medals, a record number for the U.S. at any Winter Olympics. The U.S. won twice as many medals as Canada. How many medals did Canada win?

   Solve: Check:

3. In short track speed skating, one lap is approximately 100 meters. About how many laps would a skater need to complete in order to finish a 1000 meter race?

   Solve: Check:

4. In Cross-Country Skiing, the Combined Pursuit consists of 2 different ski styles. The freestyle portion for women is 10 km, twice as long as the classic style portion. How far must skiers race using the classic style?

   Solve: Check:

A	B	D	E	G	H	I	K	L	M	N	O	R	S	T	U	W
10	12	7	21	6	5	18	24	13	14	19	9	22	11	17	8	23

___ ___ ___ ___
 1   2   3   4

Copyright © by Holt, Rinehart and Winston.
All rights reserved.

58

Holt Mathematics

Name _____ Date _____ Class _____

LESSON 2-8

Practice A
Division Equations

Match each equation in Column 1 to its solution in Column 2.

Column 1

1. $\frac{x}{4} = 5$ _____
2. $\frac{x}{3} = 8$ _____
3. $3 = \frac{x}{6}$ _____
4. $\frac{x}{7} = 7$ _____
5. $2 = \frac{x}{5}$ _____
6. $4 = \frac{x}{9}$ _____
7. $\frac{x}{1} = 5$ _____
8. $\frac{x}{3} = 5$ _____
9. $\frac{x}{3} = 9$ _____
10. $3 = \frac{x}{4}$ _____

Column 2

A. $x = 10$
B. $x = 15$
C. $x = 36$
D. $x = 49$
E. $x = 24$
F. $x = 27$
G. $x = 12$
H. $x = 20$
I. $x = 5$
J. $x = 18$

Solve each equation. Check your answers.

11. $\frac{p}{4} = 4$

12. $\frac{q}{8} = 3$

13. $7 = \frac{s}{3}$

_____ _____ _____

14. $4 = \frac{w}{9}$

15. $7 = \frac{t}{5}$

16. $\frac{m}{7} = 8$

_____ _____ _____

17. All of the students in Tim's class are divided into 4 teams of 6 students. Write a division equation using the variable *s* to show the total number of students in Tim's class.

18. There are 3 tennis balls in each can. The coach bought a total of 27 tennis balls. Write and solve a division equation using the variable *c* to find how many cans the coach bought.

_____ _____

Name _____ Date _____ Class _____

LESSON 2-8 Practice B
Division Equations

Solve each equation. Check your answers.

1. $\dfrac{s}{6} = 7$

2. $\dfrac{v}{5} = 9$

3. $12 = \dfrac{q}{7}$

4. $\dfrac{m}{2} = 16$

5. $26 = \dfrac{x}{3}$

6. $\dfrac{n}{8} = 4$

7. $\dfrac{t}{11} = 11$

8. $\dfrac{p}{7} = 10$

9. $7 = \dfrac{v}{8}$

Solve each equation.

10. $10 = \dfrac{m}{9}$

11. $\dfrac{r}{5} = 8$

12. $11 = \dfrac{x}{7}$

13. $9 = \dfrac{p}{12}$

14. $15 = \dfrac{d}{5}$

15. $\dfrac{n}{4} = 28$

16. $\dfrac{q}{2} = 134$

17. $\dfrac{u}{16} = 1$

18. $2 = \dfrac{w}{25}$

19. All the seats in the theater are divided into 6 groups. There are 35 seats in each group. Using the variable *s*, write and solve a division equation to find how many seats there are in the theater.

20. There are 16 ounces in one pound. A box of nails weighs 4 pounds. Using the variable *w*, write and solve a division equation to find how many ounces the box weighs.

Name _____ Date _____ Class _____

LESSON 2-8 Practice C
Division Equations

Solve each equation. Check your answers.

1. $\dfrac{s}{18} = 16$

2. $\dfrac{v}{24} = 3^2$

3. $\dfrac{q}{9} = 2 \cdot 7$

4. $\dfrac{m}{2^2} = 35$

5. $7 = \dfrac{x}{25}$

6. $\dfrac{n}{3^2} = 3 \cdot 4$

7. $\dfrac{t}{30} = 14$

8. $\dfrac{p}{48} = 5$

9. $\dfrac{v}{7} = 3^2$

Solve each equation.

10. $\dfrac{m}{27} = 5$

11. $\dfrac{r}{9} = 41$

12. $\dfrac{x}{(2^3)} = 7$

13. $\dfrac{p}{(16 \cdot 2)} = 6$

14. $\dfrac{d}{59} = 2$

15. $9^2 = \dfrac{n}{3}$

16. $\dfrac{q}{35} = 11$

17. $\dfrac{t}{8} = \dfrac{117}{9}$

18. $4^3 = \dfrac{w}{2^3}$

19. If $\dfrac{x}{y} = 2$, and $x + y = 6$, what are the values of x and y?

20. If $\dfrac{(3y)}{x} = 4$, and $x^2 = 36$, what are the values for x and y?

Name _____ Date _____ Class _____

LESSON 2-8 Reteach
Division Equations

You can use multiplication and division to write related number facts.

$3 \cdot 4 = 12 \qquad 12 \div 4 = 3$

Division and multiplication are inverse operations. They undo each other. So you can use multiplication to solve division equations.

To solve $\frac{x}{2} = 3$, think of a related number fact.

If $\frac{x}{2} = 3$, then $3 \cdot 2 = x$.

$3 \cdot 2 = x$
$x = 6$

Check: $\frac{x}{2} = 3$

$\frac{6}{2} \stackrel{?}{=} 3 \qquad$ *substitute*

$3 \stackrel{?}{=} 3$ ✔

$x = 6$ is the solution for $\frac{x}{2} = 3$.

Use a related number fact to solve each equation. Then check each answer.

1. $\frac{x}{2} = 4$

2. $\frac{x}{8} = 2$

3. $\frac{x}{3} = 5$

_____ _____ _____

4. $\frac{x}{5} = 1$

5. $\frac{x}{9} = 3$

6. $\frac{x}{6} = 3$

_____ _____ _____

7. $\frac{x}{8} = 4$

8. $\frac{x}{2} = 9$

9. $\frac{x}{4} = 4$

_____ _____ _____

10. $\frac{x}{5} = 4$

11. $\frac{x}{6} = 2$

12. $\frac{x}{9} = 4$

_____ _____ _____

Name _____ Date _____ Class _____

LESSON 2-8 Challenge
What Does Algebra Mean?

About 1,200 years ago, Arab people invented the branch of mathematics called *algebra*. In fact, the word *algebra* comes from the Arabic word *al-jabr*. What does that word mean?

Solve each division equation below. Then in the box at the bottom of the page, write the variable in the blank above its value. When you have solved all the equations you will have found the answer to the question.

1. $\frac{s}{4} = 6$ _____
2. $\frac{b}{3} = 5$ _____
3. $9 = \frac{p}{4}$ _____
4. $i \div 2 = 7$ _____
5. $8 = \frac{a}{6}$ _____
6. $11 = f \div 2$ _____
7. $\frac{h}{7} = 6$ _____

8. $\frac{k}{9} = 5$ _____
9. $6 = \frac{u}{2}$ _____
10. $3 = \frac{n}{7}$ _____
11. $o \div 8 = 4$ _____
12. $\frac{r}{7} = 5$ _____
13. $t \div 9 = 3$ _____
14. $8 = e \div 1$ _____

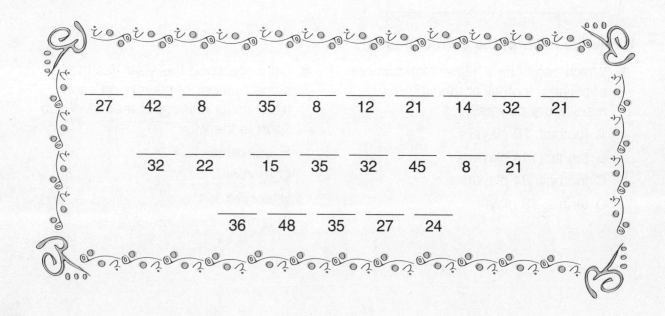

___ ___ ___ ___ ___ ___ ___ ___ ___ ___
27 42 8   35 8   12 21 14 32 21

___ ___ ___ ___ ___ ___ ___ ___
32 22 15 35 32 45 8   21

___ ___ ___ ___ ___
36 48 35 27 24

_____

Name _____ Date _____ Class _____

LESSON 2-8 Problem Solving
Division Equations

Use the table to write and solve a division equation to answer each question.

1. How many total people signed up to play soccer in Bakersville this year?

   _____

2. How many people signed up to play lacrosse this year?

   _____

3. What was the total number of people who signed up to play baseball this year?

   _____

4. Which two sports in the league have the same number of people signed up to play this year? How many people are signed up to play each of those sports?

   _____

Bakersville Sports League

Sport	Number of Teams	Players on Each Team
Baseball	7	20
Soccer	11	15
Football	8	24
Volleyball	12	9
Lacrosse	6	17
Basketball	10	10
Tennis	18	6

Circle the letter of the correct answer.

5. Which sport has a higher total number of players, football or tennis? How many more players?

   A football; 10 players
   B tennis; 144 players
   C football; 84 players
   D tennis; 18 players

6. Only one sport this year has the same number of players on each team as its number of teams. Which sport is that?

   F basketball
   G football
   H soccer
   J tennis

Name _____ Date _____ Class _____

LESSON 2-8 Reading Strategies
Follow a Sequence

Knowing that multiplication and division are **inverse operations** can help you solve division equations.

$51 \div 3 = 17$ and $17 \cdot 3 = 51$

$65 \div 5 = 13$ and $13 \cdot 5 = 65$

From these examples, you could say that multiplication **"undoes"** division. This makes sense, since multiplication and division are **opposite operations.**

Example: $s \div 18 = 5 \longrightarrow$ Read: "s divided by 18 equals 5."

Follow these steps to solve:

Step 1: Get the variable by itself. Use multiplication to "undo" division. Since s is divided by 18, you will multiply by 18.

$s \div 18 = 5$

Step 2: To keep the equation balanced, multiply the right side of the equation by 18 also.

$s \div 18 \cdot 18 = 5 \cdot 18$

Step 3: Check to verify that $s = 90$ is the solution.

$s = 90$
$s \div 18 = 5$
$90 \div 18 \stackrel{?}{=} 5$
$5 \stackrel{?}{=} 5$ ✓   90 is the solution.

Use $w \div 7 = 98$ for Exercises 1–4.

1. Write the equation in words.

_____

2. What operation is used in the equation?

_____

3. What operation will you perform on both sides of the equation to solve it?

_____

4. Write about how you would solve this problem step by step:
   $7 = z \div 20$.

_____

_____

Puzzles, Twisters & Teasers
2-8 Where Are They?

Where will you find a lobster's teeth?
<u>Hint:</u> It is called the gastric mill.

Solve each equation. Use the inverse operation to check your answers.

1. $\frac{x}{4} = 8$ **Check:**

   $4 \cdot \frac{x}{4} = 4 \cdot 8$ $\frac{32}{4} \stackrel{?}{=} 8$

   ____32____ $8 \stackrel{?}{=} 8$ ✓

2. $7 = \frac{y}{6}$ **Check:**

3. $\frac{z}{9} = 5$ **Check:**

4. $64 = \frac{m}{2}$ **Check:**

5. $\frac{s}{12} = 72$ **Check:**

6. $4 = \frac{n}{25}$ **Check:**

7. $\frac{b}{10} = 100$ **Check:**

To find the answer, use your solutions in the decoder:

A	B	C	D	E	F	G	H	I	J	L	M	N
864	12	100	702	21	111	209	1000	180	243	13	128	19

O	P	Q	R	S	T	U	V	W	X	Y	Z
45	150	33	40	32	42	500	22	34	116	8	23

In it's __S__ __T__ __O__ __M__ __A__ __C__ __H__ .
           1   2   3   4   5   6   7

Name _____ Date _____ Class _____

CHAPTER 2 Algebra Tiles
Solving Addition Equations

Holt Mathematics

Name _____ Date _____ Class _____

CHAPTER 2 **Cards for Reaching All Learners**
Solving Division Equations

1	2	3
4	5	6
7	8	9

LESSON 2-1 Practice A
Variables and Expressions

Circle the letter of the correct answer.

1. Which of the following is an algebraic expression?
   A 4 + 13
   B 10 · (3 − 2)
   C 15 ÷ 5
   D 9 − n

2. What is the variable in the expression (16 + a) · 5 − 4?
   F 16
   G a
   H 5
   J n

3. Which of these expressions is a way to rewrite the algebraic expression n ÷ 3?
   A $\frac{n}{3}$
   B n · 3
   C 3n
   D $\frac{3}{n}$

4. Which of these expressions is not a way to rewrite the algebraic expression n · 4?
   F n(4)
   G n · 4
   H $\frac{4}{n}$
   J 4n

Evaluate each expression to find the missing values in the tables.

5.
n	n + 3
1	4
2	5
5	8
7	10
10	13

6.
n	n · 2^2
2	8
3	12
5	20
7	28
8	32

7. If x = 3, what is the value of the expression 6 ÷ x?

   2

8. If x = 2, what is the value of the expression 9 − x?

   7

LESSON 2-1 Practice B
Variables and Expressions

Evaluate each expression to find the missing values in the tables.

1.
n	n + 8^2
7	71
9	73
22	86
35	99

2.
n	25 − n
20	5
5	20
18	7
9	16

3.
n	n · 7
8	56
9	63
11	77
12	84

4.
n	24 ÷ n
2	12
6	4
4	6
8	3

5.
n	n + 15
35	50
5	20
20	35
85	100

6.
n	n · 2^3
7	56
4	32
10	80
13	104

7. A car is traveling at a speed of 55 miles per hour. You want to write an algebraic expression to show how far the car will travel in a certain number of hours. What will be your constant? your variable?

   55 will be the constant, and the number of hours will be the variable.

8. Shawn evaluated the algebraic expression x ÷ 4 for x = 12 and gave an answer of 8. What was his error? What is the correct answer?

   He used subtraction instead of division. The correct answer is 3.

LESSON 2-1 Practice C
Variables and Expressions

Evaluate each expression to find the missing values in the tables.

1.

n	n ÷ 15
30	2
75	5
15	1
105	7

2.
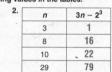
n	3n − 2^3
3	1
8	16
10	22
29	79

3.
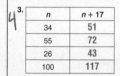
n	n + 17
34	51
55	72
26	43
100	117

4.
l	w	l × w
5	3	15
6	3	18
7	3	21
8	3	24

Evaluate each expression for the given value of the variable.

5. 5x + 2 for x = 4   6. 63 − 8z for z = 7   7. 176 ÷ p for p = 2

   22   7   88

8. $\frac{64}{v}$ − 11 for v = 4   9. 19w for w = 5   10. 98 − 5q for q = 7

   5   95   63

11. 48 ÷ n for n = 3   12. x + x + x for x = 15   13. 16 + n^2 for n = 3

    16   45   25

14. What is the next expression in the following pattern: 4n; 8n; 16n?

    32n

15. What is the next expression in the pattern x + 27; x + 24; x + 21?

    x + 18

LESSON 2-1 Reteach
Variables and Expressions

A variable is a letter or a symbol that stands for a number that can change. A constant is an amount that does not change.

A mathematical phrase that contains at least one variable is an algebraic expression. In the algebraic expression x + 5, x is a variable and 5 is a constant.

When you evaluate an algebraic expression, substitute a number for the variable and then find the value.

To evaluate the algebraic expression m − 8 for m = 12, first replace the variable m in the expression with 12.
m − 8
12 − 8
Then find the value of the expression.
12 − 8 = 4
The value of m − 8 is 4 when m = 12.

Evaluate each expression for the given value of the variable.

1. x + 5, for x = 6   2. 3p, for p = 5   3. z ÷ 4, for z = 24   4. w − 7, for w = 15

   11   15   6   8

To find the missing values in a table, use the given values of the variable.

x	4x
3	12
4	■
5	■

Think: x = 3, so 4x = 4 · 3 = 12
Think: x = 4, so 4x = 4 · 4 = 16
Think: x = 5, so 4x = 4 · 5 = 20

Evaluate each expression to find the missing values in the tables.

5.
x	x + 7
3	10
5	12
7	14

6.
y	y − 2
9	7
10	8
14	12

Holt Mathematics

LESSON 2-1 Challenge
Express Trains

Use the expression written on the side of each train's engine to find the missing values for the cars it pulls. Then choose your own value for the variable to fill in the last caboose on each train.

1. $n \div 7$: 6, 8, 4, 5; $n=42$, $n=56$, $n=28$, $n=35$
2. $2x+5$: 11, 21, 25, 15; $x=3$, $x=8$, $x=10$, $x=5$
3. $c \div 12$: 4, 2, 5, 8; $c=48$, $c=24$, $c=60$, $c=96$
4. $5p-9$: 31, 11, 46, 16; $p=8$, $p=4$, $p=11$, $p=5$
5. $7m+2m$: 45, 18, 81, 27; $m=5$, $m=2$, $m=9$, $m=3$

Possible answers are given on each caboose. Accept all answers that correctly match the chosen variable and the train's expression.

LESSON 2-1 Problem Solving
Variables and Expressions

Write the correct answer.

1. To cook 4 cups of rice, you use 8 cups of water. To cook 10 cups of rice, you use 20 cups of water. Write an expression to show how many cups of water you should use if you want to cook c cups of rice. How many cups of water should you use to cook 5 cups of rice?

   $2c$; 10 cups of water

2. Sue earns the same amount of money for each hour that she tutors students in math. In 3 hours, she earns $27. In 8 hours, she earns $72. Write an expression to show how much money Sue earns working h hours. At this rate, how much money will Sue earn if she works 12 hours?

   $9h$; $108

3. Bees are one of the fastest insects on Earth. They can fly 22 miles in 2 hours, and 55 miles in 5 hours. Write an expression to show how many miles a bee can fly in h hours. If a bee flies 4 hours at this speed, how many miles will it travel?

   $11h$; 44 miles

4. A friend asks you to think of a number, triple it, and then subtract 2. Write an algebraic expression using the variable x to describe your friend's directions. Then find the value of the expression if the number you think of is 5.

   $3x - 2$; 13

Circle the letter of the correct answer.

5. The ruble is the currency in Russia. In 2005, 1 United States dollar was worth 28 rubles. How many rubles were equivalent to 10 United States dollars?
   A 28
   B 38
   C 280
   D 2,800

6. The peso is the currency in Mexico. In 2005, 1 United States dollar was worth 10 pesos. How many pesos were equivalent to 5 United States dollars?
   F 1
   G 10
   H 15
   J 50

LESSON 2-1 Reading Strategies
Focus on Vocabulary

The word **vary** means **change**. In math, a **variable** is a letter that holds a place for numbers that change.

1. Give some examples of things that vary.

   Sample answers: population of a state; your height, age, and weight from year to year; number of people in your school; number of fish in an aquarium.

The opposite of variable is **constant**. Something that is constant never changes, such as the street number of your house or the number of inches in a foot.

2. Give some examples of things that are constant.

   Sample answers: number of wheels on a car; number of days in a week; number of ears and eyes people have; number of years in a decade.

In English, we use words in expressions such as, "see you soon" or "have a good day." In math, we use numbers and symbols to write **expressions** for other numbers.

$10 + 3$ $4 + 8 + 5$ $2(8 + 5)$

3. Write a math expression for 14.

   Sample answers: $10 + 4$ or $7 + 7$

4. Write a math expression for 25.

   Sample answers: $20 + 5$ or $10 + 15$

An **algebraic expression** is a math expression that contains a variable.

$x + 5$ $3n + 1$ $8 - w$

For Exercises 5–8, write "yes" if the expression is an algebraic expression or "no" if it is not.

5. $n + 7$ _____ yes
6. $8(y + 1)$ _____ yes
7. $6 + (10 + 5)$ _____ no
8. $4x - 1$ _____ yes

LESSON 2-1 Puzzles, Twisters & Teasers
Between Meals

What did the boat do after breakfast?

Circle each correct answer. Then put the letter above the correct answer in the box.

1. Which of the following is an algebraic expression?
   L $25x$ C $0.25 \cdot 8$ N $13 + 6 - 9$ D $\frac{3}{4}$

2. Which of the following is NOT an algebraic expression?
   T $w + 7$ M $15y$ C $24 - g$ **A** $250 - 135 \cdot 2$

3. In the following equation, which symbol is the variable?
   $x \cdot 2 = 6$
   U x H $\cdot$ O 2 A $=$ W 6

4. Evaluate the expression for the given value of the variable.
   $x \cdot 9$ or $9x$ for $x = 3$
   A 3 E 12 **N** 27

5. Solve the equation to find the value of the variable.
   $3n - 1 = 29$
   S 9 **C** 10 P 11

6. Which algebraic equation best describes the following question?
   If a kangaroo can leap 6 feet, how many leaps will it take for her to travel 72 feet?
   H $6x = 72$ J $6 + 72 = x$ T $72x = 6$

L A U N C H

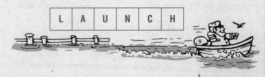

LESSON 2-2 Practice A
Translate Between Words and Math

Circle the letter of the correct answer.

1. Which of the following is the solution to an addition problem?
   A product
   (B) sum
   C plus
   D add

2. Which of the following is the solution to a subtraction problem?
   F minus
   G times
   (H) difference
   J less

3. Which word phrase represents the following expression: 5 • 3?
   (A) the product of 5 and 3
   B 5 less than 3
   C the quotient of 5 and 3
   D the sum of 5 and 3

4. Which word phrase represents the following expression: 14 ÷ n?
   F the difference of 14 and n
   G 14 more than n
   H take away n from 14
   (J) the quotient of 14 and n

Match each situation to its algebraic expression below.

A. 8 ÷ x   B. 8x   C. 8 − x   D. x + 8   E. x − 8   F. x ÷ 8

5. 8 take away x __C__
6. x divided by 8 __F__
7. the product of 8 and x __B__
8. the quotient of 8 and x __A__
9. 8 more than x __D__
10. x decreased by 8 __E__

11. Lily bought 14 beads and lost some of them. This situation is modeled by the expression 14 − x. What does x represent in the expression?
    the number of beads she lost

12. The pet store put the same number of hamsters in 6 cages. This situation is modeled by the expression 6n. What does n represent?
    the number of hamsters in each cage

LESSON 2-2 Practice B
Translate Between Words and Math

Write an expression.

1. Terry's essay has 9 more pages than Stacey's essay. If s represents the number of pages in Stacey's essay, write an expression for the number of pages in Terry's essay.
   $s + 9$

2. Let z represent the number of students in a class. Write an expression for the number of students in 3 equal groups.
   $\frac{z}{3}$

Write each phrase as a numerical or algebraic expression.

3. 24 multiplied by 3   4. n multiplied by 14   5. w added to 64
   $24 \cdot 3$ $n \cdot 14$ $64 + w$

6. the difference of 58 and 6   7. m subtracted from 100   8. the sum of 180 and 25
   $58 - 6$ $100 - m$ $180 + 25$

9. the product of 35 and x   10. the quotient of 63 and 9   11. 28 divided by p
   $35x$ $63 \div 9$ $28 \div p$

Write two phrases for each expression. Possible answers are given.

12. n + 91   n plus 91; 91 more than n
13. 35 ÷ r   35 divided by r; the quotient of 35 and r
14. 20 − s   20 minus s; s less than 20

15. Charles is 3 years older than Paul. If y represents Paul's age, what expression represents Charles's age?
    $y + 3$

16. Maya bought some pizzas for $12 each. If p represents the number of pizzas she bought, what expression shows the total amount she spent?
    $12p$

LESSON 2-2 Practice C
Translate Between Words and Math

Write each phrase as a numerical or algebraic expression.

1. the sum of 69, 140, and 300
   $69 + 140 + 300$

2. 95 less than the quotient of x and 12
   $(x \div 12) - 95$

3. 144 less than 500
   $500 - 144$

4. 22 added to the product of 14 and n
   $14n + 22$

5. The difference of 98 and p, added to 4
   $4 + (98 - p)$

6. 85 more than twice m
   $2m + 85$

Write two phrases for each expression. Possible answers are given.

7. $\frac{150}{n}$
   150 divided by n; the quotient of 150 and n

8. 79 − w
   79 take away w; the difference of 79 and w

9. 12 + 29q
   12 plus the product of 29 and q; the product of 29 and q added to 12

10. (87 − p) + 11
    the difference of 87 and p, plus 11; 11 added to 87 minus p

11. (28 ÷ x) − 6
    6 less than the quotient of 28 and x; 28 divided by x, minus 6

12. (4 + z) − 18z
    the sum of 4 and z, minus the product of 18 and z; 4 plus z, minus 18 times z

13. Mohamed bought several bottles of juice for $3 each. He paid for them all with a $20 bill. If j represents the number of bottles Mohamed bought, what expression represents the change he would receive?
    $20 - 3j$

14. A giant bamboo plant grew 18 inches per year. When Mrs. Sanchez started measuring the plant it was 5 inches tall. If y represents the number of years she measured the plant, what expression represents its height?
    $5 + 18y$

LESSON 2-2 Reteach
Translate Between Words and Math

There are key words that tell you which operations to use for mathematical expressions.

Addition (combine)	Subtraction (less)	Multiplication (put together groups of equal parts)	Division (separate into equal groups)
add	minus	product	quotient
plus	difference	times	divide
sum	subtract	multiply	
total	less than		
increased by	decreased by		
more than	take away		

You can use key words to help you translate between word phrases and mathematical phrases.

A. 3 plus 5   B. 3 times x   C. 5 less than p   D. h divided by 6
   $3 + 5$ $3x$ $p - 5$ $h \div 6$

Write each phrase as a numerical or algebraic expression.

1. 4 less than 8   2. q divided by 3   3. f minus 6   4. d multiplied by 9
   $8 - 4$ $q \div 3$ $f - 6$ $d \cdot 9$

You can use key words to write word phrases for mathematical phrases.

A. 7k
 • the product of 7 and k
 • 7 times k

B. 5 − 2
 • 5 minus 2
 • 2 less than 5

Write a phrase for each expression. Possible answers are given.

5. z ÷ 4   6. 5 • 6   7. m − 6   8. s + 3
   z divided by 4   5 times 6   6 less than m   s plus 3

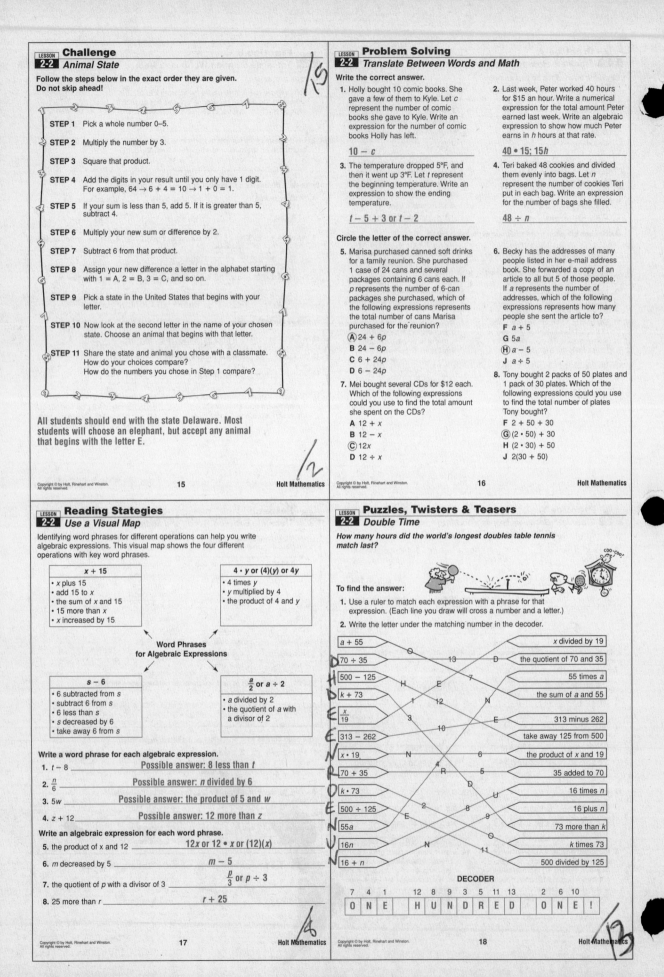

LESSON 2-3 Practice A
Translating Between Tables and Expressions

Circle the letter of the correct answer.

1. Which sentence about the table is true?

Cars	Wheels
1	4
2	8
3	12
c	4c

A The number of wheels is the number of cars plus 4.
B The number of wheels is the number of cars minus 4.
C The number of wheels is the number of cars divided by 4.
(D) The number of wheels is 4 times the number of cars.

2. Which sentence about the table is not true?

Brett's Age	Joy's Age
10	11
11	12
12	13
b	b + 1

F Joy's age is Brett's age plus 1.
G When Brett's age is b, Joy's age is b + 1.
H Add 1 to Brett's age to get Joy's age.
(J) Subtract 1 from Brett's age to get Joy's age.

Write an expression for the missing value in each table.

3.
Motorcycle	Wheels
1	2
2	4
3	6
m	2m

4.
Marbles	Bags
15	3
20	4
25	5
m	m ÷ 5

Write an expression for the sequence in the table.

5.
Position	1	2	3	4	5	n
Value of Term	3	4	5	6	7	n + 2

6. What is the value of the term in position 6 in Exercise 5? ___8___

LESSON 2-3 Practice B
Translating Between Tables and Expressions

Write an expression for the missing value in each table.

1.
Bicycles	Wheels
1	2
2	4
3	6
b	2b

2.
Ryan's Age	Mia's Age
14	7
16	9
18	11
r	r − 7

3.
Minutes	Hours
60	1
120	2
180	3
m	m ÷ 60

4.
Bags	Potatoes
3	21
4	28
5	35
b	7b

Write an expression for the sequence in each table.

5.
Position	1	2	3	4	5	n
Value of Term	3	4	5	6	7	n + 2

6.
Position	1	2	3	4	5	n
Value of Term	5	9	13	17	21	4n + 1

7. A rectangle has a width of 6 inches. The table shows the area of the rectangle for different widths. Write an expression that can be used to find the area of the rectangle when its length is l inches.

Width (in.)	Length (in.)	Area (in.²)
6	8	48
6	10	60
6	12	72
6	l	6l

LESSON 2-3 Practice C
Translating Between Tables and Expressions

Write an expression for the missing value in each table.

1.
Game	Cards
1	52
2	104
3	156
g	52g

2.
Paper Clips	Boxes
250	5
500	10
750	15
c	c ÷ 50

Write an expression for the sequence in each table.

3.
Position	1	2	3	4	5	n
Value of Term	8	13	18	23	28	5n + 3

4.
Position	1	2	3	4	5	n
Value of Term	0	1	2	3	4	n − 1

5. What is the relationship between the number of bags and the number of coins?

The number of coins is always 6 times the number of bags.

Bags	Coins
3	18
5	30
7	42
b	?

6. Look at the table below.

Position	1	2	3	4	5	n
Value of Term	1	3	5	7	9	?

What is the relationship between the positions and the values of the terms in the sequence?

The value of a term is always 1 less than the position multiplied by 2.

LESSON 2-3 Reteach
Translating Between Tables and Expressions

You can write an expression for data in a table. The expression must work for all of the data.

Cats	Legs
1	4
2	8
3	12
c	?

Think: When there is 1 cat, there are 4 legs.   4 × 1 = 4
When there are 2 cats, there are 8 legs.   4 × 2 = 8
When there are 3 cats, there are 12 legs.   4 × 3 = 12

So, when there are c cats, there are 4c legs.

You can write an expression for the sequence in a table.
Find a rule for the data in the table that works for the whole sequence.

Position	1	2	3	4	5	n
Value of Term	4	5	6	7	8	?

Step 1 Look at the value of the term in position 1.
4 is **3 more** than 1.
Step 2 Try the rule for position 2.
5 is **3 more** than 2.
Step 3 Try the rule for the rest of the positions.
6 is **3 more** than 3, 7 is **3 more** than 4, and 8 is **3 more** than 5.
So, the expression for the sequence is n + 3.

Write an expression for the missing value in each table.

1.
People	Legs
1	2
2	4
3	6
p	2p

2.
Yoko's Age	Mel's Age
9	19
10	20
11	21
y	y + 10

Write an expression for the sequence in the table.

3.
Position	1	2	3	4	5	n
Value of Term	3	6	9	12	15	3n

Holt Mathematics

LESSON 2-3 Challenge
Table Match

Match each expression with the table it belongs in.

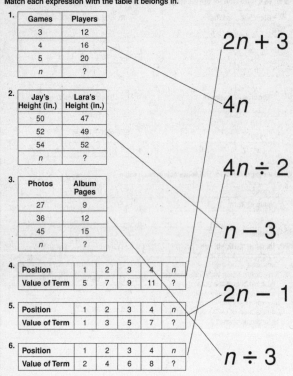

LESSON 2-3 Problem Solving
Translating Between Tables and Expressions

Use the table to write an expression for the missing value.
Then use your expression to answer the questions.

Cars Produced By Company X

Number of Years	Average Number of Cars Produced
2	2,500
5	6,250
7	8,750
10	12,500
12	15,000
14	17,500
n	$1,250n$

1. How many cars are produced on average each year?
   1,250

2. How many cars will be produced in 6 years?
   7,500

3. After how many years will there be an average production of 3,750 cars?
   3

Circle the letter of the correct answer.
Company Y produces twice as many cars as Company X.

4. How many cars does Company Y produce on average in 8 years?
   A 1,250
   B 10,000
   C 11,250
   (D) 20,000

5. How many more cars on average does Company Y produce in 4 years than Company X?
   F 2,500
   (G) 5,000
   H 6,125
   J 7,500

6. Which company produces an average of 11,250 cars in 9 years?
   (A) Company X
   B Company Y
   C both companies
   D neither company

7. How many cars are produced on average by both companies in 20 years?
   F 3,750
   G 12,500
   H 25,000
   (J) 37,500

CHAPTER Reading Strategies
2-3 Identify Relationships

When you are **related** to someone, you are connected by something in common. When you look at the positions and the values of terms in a table, they are related, too. You can find the connection, or **relationship**. Then you can write an expression for the sequence.

Position	1	2	3	4	5	n
Value of Term	10	11	12	13	14	?

Read the value of the term in the first position. Note how it is related to its position.

   1   +9   =   10
   ↑   ↑
position 1   relationship between   value of term
             position and value of term   in position 1

Check to see if the relationship works for the second position.

   2   +9   =   11
   ↑   ↑
position 2   relationship between   value of term
             position and value of term   in position 2

Check again. Use the value of the term in the third position.

   3   +9   =   12
   ↑   ↑
position 3   relationship between   value of term
             position and value of term   in position 3

So, the expression for the sequence is $n + 9$.

Use this table to answer Exercises 1–6.

Position	1	2	3	4	5	n
Value of Term	5	10	15	20	25	?

1. How do you go from position 1 to the value of its term, 5? **add 4**
2. Try the relationship for the next term. Can you add 4 to 2 and get 10? **no**
   Does $n + 4$ work? **no**
3. Try another relationship for position 1 and its term.
   How else can you go from 1 to 5? **multiply by 5**
4. Try this relationship for the next term. Can you multiply 2 by 5 and get 10? **yes**
5. Check again by using the value of the term in the third position.
   Can you multiply 3 by 5 and get 15? **yes**
6. What is the expression for the sequence in the table? **$5n$**

LESSON 2-3 Puzzles, Twisters & Teasers
Weather Forecast

What kind of animal can forecast the weather?

Use the table below to answer Exercises 1– 8. Circle the correct answer. Then put the letter above the correct answer on the lines.

Position	1	2	3	4	5	n
Value of Term	4	7	10	13	16	?

1. What is the value of the 6th term?
   K 16   A 17   B 18   (R) 19

2. What is the value of the 7th term?
   P 18   J 20   (E) 22   O 24

3. What is the value of the 8th term?
   L 20   (I) 25   F 30   D 35

4. What is the value of the 9th term?
   X 21   C 22   (N) 28   H 38

5. What is the value of the 10th term?
   (D) 31   S 41   W 51   U 61

6. What is the value of the 11th term?
   L 24   Q 30   (E) 34   Z 43

7. What is the value of the 12th term?
   B 34   G 35   M 36   (E) 37

8. What is the value of the 13th term?
   (R) 40   T 42   A 44   Y 46

 R   E   I   N   D   E   E   R
 1.   2.   3.   4.   5.   6.   7.   8.

74 Holt Mathematics

Practice A
2-4 Equations and Their Solutions

Determine whether the given value of the variable is a solution.

1. $x + 1 = 5$ for $x = 4$ __Yes__
2. $13 - w = 10$ for $w = 2$ __No__
3. $2 \cdot v = 12$ for $v = 10$ __No__
4. $14 \div p = 2$ for $p = 7$ __Yes__
5. $8 + w = 11$ for $w = 3$ __Yes__
6. $4t = 20$ for $t = 5$ __Yes__
7. $\frac{12}{s} = 4$ for $s = 3$ __Yes__
8. $6 + d = 15$ for $d = 8$ __No__

Circle the letter of the equation that each given solution makes true.

9. $x = 5$
   - **(A)** $2 + x = 7$
   - **B** $9 - x = 3$
   - **C** $3 \cdot x = 18$
   - **D** $20 \div x = 2$

10. $g = 7$
    - **F** $9g = 16$
    - **(G)** $8 - g = 1$
    - **H** $11 + g = 17$
    - **J** $g \div 1 = 1$

11. $y = 2$
    - **A** $5 + y = 8$
    - **B** $7 - y = 1$
    - **(C)** $3 \cdot y = 6$
    - **D** $10 \div y = 20$

12. $m = 9$
    - **F** $10 + m = 20$
    - **G** $m - 4 = 13$
    - **H** $7 \cdot m = 36$
    - **(J)** $18 \div m = 2$

13. $z = 4$
    - **(A)** $5z = 20$
    - **B** $12 \div z = 4$
    - **C** $z - 3 = 7$
    - **D** $z + 8 = 4$

14. $a = 8$
    - **F** $2a = 10$
    - **(G)** $a + 12 = 20$
    - **H** $a \div 4 = 4$
    - **J** $12 - a = 6$

15. Emanuel put 12 marbles on one pan of a scale. On the other pan, he put 4 marbles, then he added 8 more marbles to that side. Each of the marbles weighs 1 ounce. Is the scale balanced? Explain.

    __Yes; because 8 + 4 = 12__

16. Bill and Rhonda have the same amount of money. Bill has $13. Rhonda has one $5 bill, three $1 bills, and one other bill. Is it a $1 bill or a $5 bill? Explain.

    __a $5 bill; because__
    __13 = 5 + 3 + 5__

Practice B
2-4 Equations and Their Solutions

Determine whether the given value of the variable is a solution.

1. $9 + x = 21$ for $x = 11$ __No__
2. $n - 12 = 5$ for $n = 17$ __Yes__
3. $25 \cdot r = 75$ for $r = 3$ __Yes__
4. $72 \div q = 8$ for $q = 9$ __Yes__
5. $28 + c = 43$ for $c = 15$ __Yes__
6. $u \div 11 = 10$ for $u = 111$ __No__
7. $\frac{k}{8} = 4$ for $k = 24$ __No__
8. $16x = 48$ for $x = 3$ __Yes__
9. $73 - f = 29$ for $f = 54$ __No__
10. $67 - j = 25$ for $j = 42$ __Yes__
11. $39 \div v = 13$ for $v = 3$ __Yes__
12. $88 + d = 100$ for $d = 2$ __No__
13. $14p = 20$ for $p = 5$ __No__
14. $6w = 30$ for $w = 5$ __Yes__
15. $7 + x = 70$ for $x = 10$ __No__
16. $6 \cdot n = 174$ for $n = 29$ __Yes__

Replace each ? with a number that makes the equation correct.

17. $5 + 1 = 2 + \boxed{?}$ __4__
18. $10 - \boxed{?} = 12 - 7$ __5__
19. $\boxed{?} \cdot 3 = 2 \cdot 9$ __6__
20. $28 \div 4 = 14 \div \boxed{?}$ __2__
21. $\boxed{?} + 8 = 6 + 3$ __1__
22. $12 \cdot 0 = \boxed{?} \cdot 15$ __0__

23. Carla had $15. After she bought lunch, she had $8 left. Write an equation using the variable x to model this situation. What does your variable represent?

    __$15 - x = 8$; x = the amount__
    __she spent on lunch__

24. Seventy-two people signed up for the soccer league. After the players were evenly divided into teams, there were 6 teams in the league. Write an equation to model this situation using the variable x.

    __$72 \div x = 6$__

Practice C
2-4 Equations and Their Solutions

For each equation, determine whether the given value of the variable is a solution.

1. $2d = 24$ for $d = 8$ __No__
2. $15 \div p = 6$ for $p = 3$ __No__
3. $x^2 = 25$ for $x = 5$ __Yes__
4. $135 \div x = 9$ for $x = 15$ __Yes__
5. $7 + t - 4 = 22$ for $t = 20$ __No__
6. $\frac{4s}{8} = 6$ for $s = 12$ __Yes__
7. $\frac{u}{13} = 5$ for $u = 65$ __Yes__
8. $2x + 3x = 60$ for $x = 12$ __Yes__
9. $2(100 - k) = 64$ for $k = 36$ __No__
10. $(69 \div m) - 5 = 18$ for $m = 3$ __Yes__
11. $11 \div w = 11$ for $w = 1$ __Yes__
12. $576 + n = 1{,}000$ for $n = 524$ __No__
13. $(15 \cdot 6)y = 450$ for $y = 5$ __Yes__
14. $18c \div 2 = 89$ for $c = 11$ __No__
15. $6^2 - r = 20$ for $r = 26$ __No__
16. $x^5 = 32$ for $x = 2$ __Yes__

Replace each ? with a number that makes the equation correct.

17. $28 + 11 = 22 + \boxed{?}$ __17__
18. $19 - \boxed{?} = 75 - 60$ __4__
19. $\boxed{?} \cdot 12 = 21 \cdot 4$ __7__
20. $54 \div 6 = 108 \div \boxed{?}$ __12__
21. $\boxed{?} + 87 = 46 + 59$ __18__
22. $3^2 \cdot 3 = \boxed{?} \cdot 3^3$ __1__

23. Mr. Yakima teaches 4 science classes, with the same number of students in each class. Of those students, 80 are sixth graders, and 40 are fifth graders. Write an equation to model this situation using the variable n. What does your variable represent?

    __$4n = 80 + 40$; n = the number__
    __of students in each class__

24. Mary put new tiles on her kitchen floor. The floor measures 6 feet long by 5 feet wide. She used 10 tiles to cover the entire floor. Each tile was 3 feet long. Write an equation using the variable x to model this situation. What does x represent in the equation?

    __$(6 \cdot 5) = 3x \cdot 10$; x = the width__
    __of each tile__

Reteach
2-4 Equations and Their Solutions

An equation is a mathematical sentence that says that two quantities are equal.

Some equations contain variables. A solution for an equation is a value for a variable that makes the statement true.

You can write related facts using addition and subtraction.
$7 + 6 = 13 \qquad 13 - 6 = 7$

You can write related facts using multiplication and division.
$3 \cdot 4 = 12 \qquad 12 \div 4 = 3$

You can use related facts to find solutions for equations. If the related fact matches the value for the variable, then that value is a solution.

A. $x + 5 = 9$, when $x = 3$
Think: $9 - 5 = x$
$x = 4$
$3 \neq 4$
So $x = 3$ is not a solution of $x + 5 = 9$.

B. $x - 7 = 5$, when $x = 12$
Think: $5 + 7 = x$
$x = 12$
$12 = 12$
So $x = 12$ is a solution of $x - 7 = 5$.

C. $2x = 14$, when $x = 9$
Think: $14 \div 2 = x$
$x = 7$
$9 \neq 7$
So $x = 9$ is not a solution for $2x = 14$.

D. $x \div 5 = 3$, when $x = 15$
Think: $3 \cdot 5 = x$
$x = 15$
$15 = 15$
So $x = 15$ is a solution for $x \div 5 = 3$.

Use related facts to determine whether the given value is a solution for each equation.

1. $x + 6 = 14$, when $x = 8$ __yes__
2. $s \div 4 = 5$, when $s = 24$ __no__
3. $g - 3 = 7$, when $g = 11$ __no__
4. $3a = 18$, when $a = 6$ __yes__
5. $26 = y - 9$, when $y = 35$ __yes__
6. $b \cdot 5 = 20$, when $b = 3$ __no__
7. $15 = v \div 3$, when $v = 45$ __yes__
8. $11 = p + 6$, when $p = 5$ __yes__
9. $6k = 78$, when $k = 12$ __no__

Challenge
2-4 Keep It Balanced

Study the scales below. Then circle the solution below each scale that will keep it balanced.

1. $a + 17 = 25$

    (a = 8)   a = 9

2. $6 = 24 \div x$

    x = 3   (x = 4)

3. $39 = 3n$

    n = 12   (n = 13)

4. $46 - c = 27$

    (c = 19)   c = 29

5. $p \div 7 = 9$

    p = 49   (p = 63)

6. $52 + m = 100$

    (m = 48)   m = 58

Problem Solving
2-4 Equations and Their Solutions

Use the table to write and solve an equation to answer each question. Then use your answers to complete the table.

1. A hippopotamus can stay underwater 3 times as long as a sea otter can. How long can a sea otter stay underwater?

   $3x = 15; x = 5;$
   5 minutes

2. A seal can stay underwater 10 minutes longer than a muskrat can. How long can a muskrat stay underwater?

   $x + 10 = 22; x = 12;$
   12 minutes

3. A sperm whale can stay underwater 7 times longer than a sea cow can. How long can a sperm whale stay underwater?

   $x \div 7 = 16; x = 112;$
   112 minutes

How Many Minutes Can Mammals Stay Underwater?	
Hippopotamus	15
Human	1
Muskrat	12
Platypus	10
Polar bear	2
Sea cow	16
Sea otter	5
Seal	22
Sperm whale	112

Circle the letter of the correct answer.

4. The difference between the time a platypus and a polar bear can stay underwater is 8 minutes. How long can a polar bear stay underwater?
   A 1 minute
   (B) 2 minutes
   C 3 minutes
   D 5 minutes

5. When you divide the amount of time any of the animals in the table can stay underwater by itself, the answer is always the amount of time the average human can stay underwater. How long can the average human stay underwater?
   F 6 minutes
   G 4 minutes
   H 2 minutes
   (J) 1 minute

Reading Strategies
2-4 Focus on Vocabulary

You can see the word **equal** in **equation**. In math, an equation indicates that two quantities are equal, or the same. The **= sign** in an equation separates one quantity from the other. The value on each side of the = sign is the same.

Look at the equations below. Notice how the value on each side of the = sign is the same for each equation:

$5 + 7 = 8 + 4 \qquad 19 - 7 = 12 \qquad 42 = 3 \cdot 14$

If an equation contains a variable, and the variable is replaced by a value that keeps the equation equal, that value is called a **solution** of the equation.

Examples:
$y \div 4 = 15 \qquad y \div 4 = 15$
$80 \div 4 \neq 15 \qquad 60 \div 4 = 15$
"80 divided by 4 is not equal to 15."   "60 divided by 4 is equal to 15."

Which are equations? Write the correct sign, = or ≠.

1. $7 + 23 \;\boxed{=}\; 9 + 21$
2. $35 + 15 \;\boxed{\neq}\; 45$
3. $28 - 7 \;\boxed{=}\; 15 + 6$

Replace the given value for the variable. Is it a solution?

4. $d + 28 = 45$ for $d = 17$ — $d = 17$ is a solution.
5. $84 \div s = 28$ for $s = 3$ — $s = 3$ is a solution.
6. $17 = 56 - t$ for $t = 40$ — $t = 40$ is not a solution.
7. $86 = 4w$ for $w = 24$ — $w = 24$ is not a solution.

Puzzles, Twisters & Teasers
2-4 Space Fact

On July 4, 1997, what did the Pathfinder spacecraft do?

For each equation, determine whether the given value of the variable is a solution. If it is a solution, circle =. If it is not a solution, circle ≠. Put the letter above the correct answer in the box.

1. $117 = 97 + n$ for $n = 10$ A(=) I≠   **I**
2. $96 \div x = 8$ for $x = 12$ T=   S(≠)   **T**
3. $132 \div m = 12$ for $m = 12$ K=   L(≠)   **L**
4. $k + 18 = 63$ for $k = 44$ E=   A(≠)   **A**
5. $35 \div s = 7$ for $s = 5$ N(=)   M≠   **N**
6. $44 = t - 55$ for $t = 88$ J=   D(≠)   **D**
7. $a - 6 = 36$ for $a = 42$ E(=)   A≠   **E**
8. $u \cdot 7 = 72$ for $u = 8$ P=   D(≠)   **D**
9. $b + 21 = 28$ for $b = 8$ I=   O(≠)   **O**
10. $92 - 28 = 8y$ for $y = 8$ N(=)   F≠   **N**
11. $6x = 54$ for $x = 9$ M(=)   P≠   **M**
12. $149 = 79 + 2y$ for $y = 35$ A(=)   E≠   **A**
13. $17w - 50 = 0$ for $w = 3$ T=   R(≠)   **R**
14. $25g = 25{,}000$ for $g = 1{,}000$ S(=)   H≠   **S**

Practice A
2-5 Addition Equations

Match each equation in Column 1 to its solution in Column 2.

Column 1		Column 2
1. $5 + x = 8$	**B**	A. $x = 5$
2. $12 + x = 12$	**I**	B. $x = 3$
3. $x + 11 = 15$	**G**	C. $x = 11$
4. $x + 9 = 20$	**C**	D. $x = 9$
5. $8 + x = 13$	**A**	E. $x = 7$
6. $6 + x = 14$	**H**	F. $x = 2$
7. $2 + x = 11$	**D**	G. $x = 4$
8. $x + 29 = 30$	**J**	H. $x = 8$
9. $3 + x = 10$	**E**	I. $x = 0$
10. $x + 17 = 19$	**F**	J. $x = 1$

Solve each equation. Check your answers.

11. $p + 8 = 14$
    $p = 6; 6 + 8 = 14$

12. $q + 10 = 13$
    $q = 3; 3 + 10 = 13$

13. $7 + s = 15$
    $s = 8; 7 + 8 = 15$

14. $4 + w = 11$
    $w = 7; 4 + 7 = 11$

15. $t + 12 = 15$
    $t = 3; 3 + 12 = 15$

16. $9 + m = 14$
    $m = 5; 9 + 5 = 14$

17. Phyllis has 6 yards of material. She needs 8 yards to make curtains. This situation is modeled by the equation $6 + x = 8$, where x is the amount of material she needs to buy. How much more material does she need to buy to make the curtains?

    2 yards

18. Emma paid $26 in all for a hammer and a screwdriver. The hammer cost $10. Write an addition equation using the variable n to show how much she spent on the screwdriver.

    $10 + n = 26$

Practice B
2-5 Addition Equations

Solve each equation. Check your answers.

1. $s + 3 = 23$
   $s = 20; 20 + 3 = 23$

2. $v + 10 = 49$
   $v = 39; 39 + 10 = 49$

3. $q + 9 = 16$
   $q = 7; 7 + 9 = 16$

4. $81 + m = 90$
   $m = 9; 81 + 9 = 90$

5. $38 + x = 44$
   $x = 6; 38 + 6 = 44$

6. $28 + n = 65$
   $n = 37; 28 + 37 = 65$

7. $t + 31 = 50$
   $t = 19; 19 + 31 = 50$

8. $25 + p = 39$
   $p = 14; 25 + 14 = 39$

9. $19 + v = 24$
   $v = 5; 19 + 5 = 24$

Solve each equation.

10. $m + 8 = 17$
    $m = 9$

11. $r + 14 = 20$
    $r = 6$

12. $25 + x = 32$
    $x = 7$

13. $47 + p = 55$
    $p = 8$

14. $19 + d = 27$
    $d = 8$

15. $13 + n = 26$
    $n = 13$

16. $q + 12 = 19$
    $q = 7$

17. $34 + f = 43$
    $f = 9$

18. $52 + w = 68$
    $w = 16$

19. Kenya bought 28 beads, and Nancy bought 25 beads. It takes 35 beads to make a necklace. Write and solve two addition equations to find how many more beads they each need to make a necklace.

    Kenya: $28 + b = 35$;
    $b = 7$ beads;
    Nancy: $25 + b = 35$;
    $b = 10$ beads

20. During a sales trip, Mr. Jones drove 15 miles east from Brownsville to Carlton. Then he drove several more miles east from Carlton to Sun City. The distance from Brownsville to Sun City is 35 miles. Write and solve an addition equation to find how many miles it is from Carlton to Sun City.

    $15 + m = 35$;
    $m = 20$ miles

Practice C
2-5 Addition Equations

Solve each equation. Check your answers.

1. $s + 67 = 101$
   $s = 34; 34 + 67 = 101$

2. $v + 13 = 28 - 5$
   $v = 10; 10 + 13 = 28 - 5$

3. $29 + q + 18 = 51$
   $q = 4; 29 + 4 + 18 = 51$

4. $4^2 + m = 35$
   $m = 19; 4^2 + 19 = 35$

5. $78 + x = 121 - 4$
   $x = 39; 78 + 39 = 121 - 4$

6. $6 + n = 28 - 9$
   $n = 13; 6 + 13 = 28 - 9$

7. $t + 1,906 = 2,000$
   $t = 94; 94 + 1,906 = 2,000$

8. $41 + p + 16 = 99$
   $p = 42; 41 + 42 + 16 = 99$

9. $201 + v + 30 = 249$
   $v = 18; 201 + 18 + 30 = 249$

Solve each equation.

10. $m + 38 = 90$
    $m = 52$

11. $12 + r + 17 = 60$
    $r = 31$

12. $115 + x = 320$
    $x = 205$

13. $57 + p = 63 + 18$
    $p = 24$

14. $2^3 + d = 21$
    $d = 13$

15. $15 + n = 8^2$
    $n = 49$

16. $q + 6 + 77 = 100$
    $q = 17$

17. $13,687 + t = 20,441$
    $t = 6,754$

18. $25 + w + 2 = 37$
    $w = 10$

19. Alice bought a round-trip ticket to fly from Baltimore to Chicago on SuperAir for $250. That was $16 more than she would have paid on Jet Airlines, which only offered a one-way fair. How much did Jet Airlines charge to fly from Baltimore to Chicago?

    $234

20. Together, Ken, Judy, Sam, and Ali worked a total of 83 hours last week. Ken worked 3 more hours than Judy. Ali worked 5 more hours than Ken. Judy worked 4 more hours than Sam. If Sam worked 15 hours, how many hours did Ken, Judy, and Ali each work?

    Ken: 22 hours; Judy: 19 hours;
    Ali: 27 hours

Reteach
2-5 Addition Equations

To solve an equation, you need to get the variable alone on one side of the equal sign.

You can use tiles to help you solve addition equations.

Subtraction undoes addition, so you can use subtraction to solve addition equations.

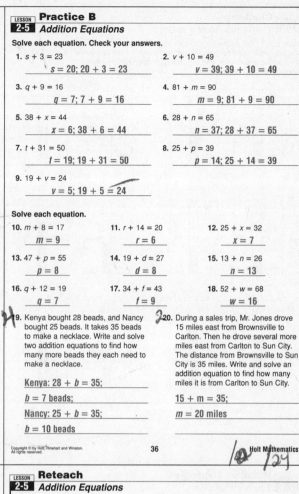

One positive tile and one negative tile is called a zero pair because together they have a value of zero.

To solve $x + 3 = 5$, first use tiles to model the equation.

Next, add enough subtraction tiles to get the variable alone. Then add the same number of subtraction tiles to the other side of the equal sign.

Then remove the greatest possible number of zero pairs from each side of the equal sign.

Check: $x + 3 = 5$
$2 + 3 \stackrel{?}{=} 5$
$5 \stackrel{?}{=} 5$ ✓

The remaining tiles represent the solution.
$x = 2$

Use tiles to solve each equation. Then check each answer.

1. $x + 8 = 11$
   $x = 3; 3 + 8 = 11$

2. $x + 4 = 9$
   $x = 5; 5 + 4 = 9$

3. $x + 7 = 13$
   $x = 6; 6 + 7 = 13$

77 Holt Mathematics

LESSON 2-5 Challenge
The Temperature's Rising

Each pair of thermometers shows a beginning temperature on the left and an ending temperature on the right. Write and solve an addition equation to find the change in temperature shown on each pair of thermometers.

1. $55 + x = 60; x = 5°F$
2. $20 + x = 30; x = 10°F$
3. $23 + x = 27; x = 4°F$
4. $8 + x = 15; x = 7°F$
5. $87 + x = 98; x = 11°F$
6. $14 + x = 27; x = 13°F$

LESSON 2-5 Problem Solving
Addition Equations

Use the bar graph and addition equations to answer the questions.

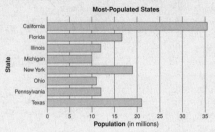

Most-Populated States

1. How many more people live in California than in New York?
$19 + x = 36; x = 17$;
17 million people

2. How many more people live in Ohio than in Michigan?
$10 + x = 11; x = 1$;
1 million people

3. How many more people live in Florida than in Illinois?
$12 + x = 17; x = 5$;
5 million people

4. How many more people live in Texas than in Pennsylvania?
$12 + x = 21; x = 9$;
9 million people

Circle the letter of the correct answer.

5. Which two states' populations are used in the equation $12 + x = 22$?
   A Pennsylvania and Texas
   B Ohio and Florida
   C Michigan and Illinois
   D Illinois and Pennsylvania

6. What is the value of x in the equation in Exercise 5?
   F 0
   G 10
   H 12
   J 24

7. In 2003, the total population of the United States was 292 million. How many of those people did not live in one of the states shown on the graph?
   A 416 million **C 154 million**
   B 73 million   D 292 million

8. The combined population of Ohio and one other state is the same as the population of Texas. What is that state?
   F California
   G Florida
   H Michigan
   J Pennsylvania

LESSON 2-5 Reading Strategies
Use a Visual Cue

An equation is like a balance scale. The value on the right side of the balance scale or equation is equal to the value on the left side of the balance scale or equation.

A balanced scale also helps you to picture a balanced equation:

Step 1: To find the value of f, the variable needs to be by itself on one side of the equation. So 32 must be subtracted from the left side of the equation.

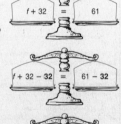

Step 2: To keep the scale balanced, subtract 32 from the right side of the equation as well.

Step 3: Check to verify that $f = 29$ is the solution.
$f + 32 = 61$
$29 + 32 \stackrel{?}{=} 61$
$61 \stackrel{?}{=} 61$ ✓   29 is the solution.

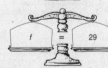

To get the variable by itself in an addition equation, subtract the same value from both sides of the equation.

Use $z + 16 = 42$ to answer Exercises 1–4.

1. On which side of the equation is the variable? __the left side__
2. What will you do to get the variable by itself? __subtract 16__
3. What must you do to the other side of the equation to keep it balanced? __subtract 16__
4. What is the value of z? __$z = 26$__

LESSON 2-5 Puzzles, Twisters & Teasers
Numbers "r" Us

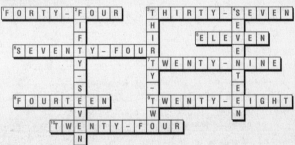

Across: 1 FORTY-FOUR, 3 THIRTY-SEVEN, 5 ELEVEN, 6 SEVENTY-FOUR, 7 TWENTY-NINE, 8 FOURTEEN, 9 TWENTY-EIGHT, 10 TWENTY-FOUR

Solve the equations. Check your work.
Write the number words of the solution in the puzzle.

Across
1. $31 + n = 75$
3. $75 = m + 38$
5. $29 = 18 + c$
6. $77 + t = 151$
7. $w + 52 = 81$
8. $57 + z = 71$
9. $a + 32 = 60$
10. $40 = p + 16$

Down
2. $b + 96 = 153$
3. $x + 15 = 47$
4. $z + 89 = 106$

LESSON 2-6 Practice A
Subtraction Equations

Match each equation in Column 1 to its solution in Column 2.

Column 1		Column 2
1. $x - 3 = 1$	__G__	A. $x = 5$
2. $3 = x - 5$	__H__	B. $x = 3$
3. $x - 1 = 6$	__E__	C. $x = 10$
4. $2 = x - 4$	__J__	D. $x = 9$
5. $x - 2 = 3$	__A__	E. $x = 7$
6. $4 = x - 6$	__C__	F. $x = 11$
7. $x - 8 = 1$	__D__	G. $x = 4$
8. $7 = x - 5$	__I__	H. $x = 8$
9. $x - 1 = 2$	__B__	I. $x = 12$
10. $9 = x - 2$	__F__	J. $x = 6$

Solve each equation. Check your answers.

11. $p - 6 = 4$
    $p = 10; 10 - 6 = 4$

12. $q - 3 = 9$
    $q = 12; 12 - 3 = 9$

13. $19 = s - 2$
    $s = 21; 19 = 21 - 2$

14. $7 = w - 4$
    $w = 11; 7 = 11 - 4$

15. $5 = t - 8$
    $t = 13; 5 = 13 - 8$

16. $m - 11 = 6$
    $m = 17; 17 - 11 = 6$

17. A box of pencils costs $3. Brian got $2 change after paying for one box. Write a subtraction equation using the variable x to show how much Brian gave the cashier.
    $x - 3 = 2$

18. After dropping 5°F, the temperature was 25°F. Write a subtraction equation using the variable t to show what the starting temperature was.
    $t - 5 = 25$

LESSON 2-6 Practice B
Subtraction Equations

Solve each equation. Check your answers.

1. $s - 8 = 12$
   $s = 20; 20 - 8 = 12$

2. $v - 11 = 7$
   $v = 18; 18 - 11 = 7$

3. $9 = q - 5$
   $q = 14; 9 = 14 - 5$

4. $m - 21 = 5$
   $m = 26; 26 - 21 = 5$

5. $34 = x - 12$
   $x = 46; 34 = 46 - 12$

6. $n - 45 = 45$
   $n = 90; 90 - 45 = 45$

7. $t - 19 = 9$
   $t = 28; 28 - 19 = 9$

8. $p - 6 = 27$
   $p = 33; 33 - 6 = 27$

9. $15 = v - 68$
   $v = 83; 15 = 83 - 68$

Solve each equation.

10. $7 = m - 5$
    $m = 12$

11. $r - 10 = 22$
    $r = 32$

12. $16 = x - 4$
    $x = 20$

13. $40 = p - 11$
    $p = 51$

14. $28 = d - 6$
    $d = 34$

15. $n - 9 = 42$
    $n = 51$

16. $q - 85 = 8$
    $q = 93$

17. $f - 13 = 18$
    $f = 31$

18. $47 = w - 38$
    $w = 85$

19. Ted took 17 pictures at the aquarium. He now has 7 pictures left on the roll. Write and solve a subtraction equation to find out how many photos Ted had when he went to the aquarium.
    $x - 17 = 7; x = 24$ photos

20. Ted bought a dolphin poster for $12. He now has $5. Write and solve a subtraction equation to find out how much money Ted took to the aquarium.
    $x - 12 = 5; x = 17

LESSON 2-6 Practice C
Subtraction Equations

Solve each equation. Check your answers.

1. $s - 57 = 38$
   $s = 95; 95 - 57 = 38$

2. $v - 16 = 12 + 6$
   $v = 34; 34 - 16 = 12 + 6$

3. $q - 18 - 5 = 20$
   $q = 43; 43 - 18 - 5 = 20$

4. $m - 3^2 = 15$
   $m = 24; 24 - 3^2 = 15$

5. $159 = x - 78$
   $x = 237; 159 = 237 - 78$

6. $n - 4^2 = 4$
   $n = 20; 20 - 4^2 = 4$

7. $t - 4,360 = 1,804$
   $t = 6,164; 6,164 - 4,360 = 1,804$

8. $p - 63 - 14 = 99$
   $p = 176; 176 - 63 - 14 = 99$

9. $v - 50 = 14 + 9$
   $v = 73; 73 - 50 = 14 + 9$

Solve each equation.

10. $m - 79 = 12$
    $m = 91$

11. $r - 109 = 65$
    $r = 174$

12. $x - 58 = 370$
    $x = 428$

13. $p - 16 = 7 + 6$
    $p = 29$

14. $d - 2^4 = 20$
    $d = 36$

15. $7^2 = n - 11$
    $n = 60$

16. $q - 12 - 140 = 15$
    $q = 167$

17. $t - 18,620 = 19,000$
    $t = 37,620$

18. $w - 3^2 = 16 + 2$
    $w = 27$

19. If $x - y = 6$, and $x = y + y$, what are the values for x and y?
    $x = 12$ and $y = 6$

20. If $x + 4 = y - 2$, and $y = x + x + x$, what are the values for x and y?
    $x = 3$ and $y = 9$

LESSON 2-6 Reteach
Subtraction Equations

To solve an equation, you need to get the variable alone on one side of the equal sign.

You can use tiles to help you solve subtraction equations.

Addition undoes subtraction, so you can use addition to solve subtraction equations.

variable   add 1   subtract 1   add 1   subtract 1   ← zero pair

One positive tile and one negative tile is called a zero pair because together they have a value of zero.

To solve $x - 4 = 2$, first use tiles to model the equation.

$x - 4 = 2$

Next, add enough addition tiles to get the variable alone. Then add the same number of addition tiles to the other side of the equal sign.

$x - 4 + 4 = 2 + 4$

Then remove the greatest possible number of zero pairs from each side of the equal sign.

$x = 6$

Check: $x - 4 = 2$
$6 - 4 \stackrel{?}{=} 2$
$2 \stackrel{?}{=} 2$ ✓

The remaining tiles represent the solution.
$x = 6$

Use tiles to solve each equation. Then check each answer.

1. $x - 5 = 3$
   $x = 8; 8 - 5 = 3$

2. $x - 2 = 5$
   $x = 7; 7 - 2 = 5$

3. $x - 6 = 4$
   $x = 10; 10 - 6 = 4$

4. $x - 8 = 1$
   $x = 9; 9 - 8 = 1$

5. $x - 3 = 9$
   $x = 12; 12 - 3 = 9$

6. $x - 7 = 3$
   $x = 10; 10 - 7 = 3$

Challenge
2-6 The Price Is Right

Each of the grocery items on this page has a different price—$1, $2, $3, $4, or $5. Use logic and the subtraction equations below to figure out the price of each item. Then write the correct price on each item's price tag.

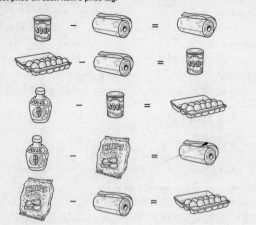

Problem Solving
2-6 Subtraction Equations

Write and solve subtraction equations to answer the questions.

1. Dr. Felix Hoffman invented aspirin in 1899. That was 29 years before Alexander Fleming invented penicillin. When was penicillin invented?

   $x - 29 = 1899;$
   $x = 1928;$ in 1928

2. Kimberly was born on February 2. That is 10 days earlier than Kent's birthday. When is Kent's birthday?

   $x - 10 = 2;$
   $x = 12;$ February 12

3. Kansas and North Dakota are the top wheat-producing states. In 2000, North Dakota produced 314 million bushels of wheat, which was 34 million bushels less than Kansas produced. How much wheat did Kansas farmers grow in 2000?

   $x - 34 = 314;$
   $x = 348;$ 348 million bushels

4. Scientists assign every element an atomic number, which is the number of protons in the nucleus of that element. The atomic number of silver is 47, which is 32 less than the atomic number of gold. How many protons are in the nucleus of gold?

   $x - 32 = 47;$
   $x = 79;$ 79 protons

Circle the letter of the correct answer.

5. The spine-tailed swift and the frigate bird are the two fastest birds on earth. A frigate bird can fly 95 miles per hour, which is 11 miles per hour slower than a spine-tailed swift. How fast can a spine-tailed swift fly?
   A 84 miles per hour
   B 101 miles per hour
   C 106 miles per hour
   D 116 miles per hour

6. The Green Bay Packers and the Kansas City Chiefs played in the first Super Bowl in 1967. The Chiefs lost by 25 points, with a final score of 10. How many points did the Packers score in the first Super Bowl?
   F 35
   G 25
   H 15
   J 0

7. The Rocky Mountains extend 3,750 miles across North America. That is 750 miles shorter than the Andes Mountains in South America. How long are the Andes Mountains?
   A 3,000 miles   C 180 miles
   B 5 miles   **D** 4,500 miles

8. When the United States took its first census in 1790, only 4 million people lived here. That was 288 million fewer people than the population in 2003. What was the population of the United States in 2003?
   F 292 million   H 69 million
   G 284 million   J 1,108 million

Reading Strategies
2-6 Use a Visual Cue

You can picture balanced scales and follow similar steps to solve subtraction equations.

Picture balanced scales for this equation.

Step 1: To find the value of b, get b by itself on the left side of the equation. So add 17 to the left side of the equation.

Step 2: To keep the equation balanced, add 17 to the right side of the equation as well.

Step 3: Check to verify that $b = 82$ is the solution.
   $b - 17 = 65$
   $82 - 17 \stackrel{?}{=} 65$
   $65 \stackrel{?}{=} 65$ ✓   82 is the solution.

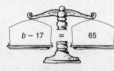

To get the variable by itself in a subtraction equation, add the same value to both sides of the equation.

Use $t - 18 = 53$ to answer Exercises 1–4.

1. On which side of the equation is the variable? _the left side_
2. What will you do to get the variable by itself? _add 18_
3. What must you do to the other side of the equation to keep it balanced? _add 18_
4. What is the value of t? $t = 71$

Puzzles, Twisters & Teasers
2-6 Opposites Attract

Why are addition and subtraction like poems?

Solve the following equations to decode the answer.

1. $n - 16 = 12$ $n = $ __28__
2. $10 = y - 13$ $y = $ __23__
3. $s - 6 = 2$ $s = $ __8__
4. $26 = i - 14$ $i = $ __40__
5. $g - 82 = 93$ $g = $ __175__
6. $37 = a - 5$ $a = $ __42__
7. $r - 7 = 23$ $r = $ __30__
8. $64 = v - 27$ $v = $ __91__
9. $46 = b - 31$ $b = $ __77__
10. $e - 5 = 4$ $e = $ __9__

They are _i_ _n_ _v_ _e_ _r_ _s_ _e_ _s_ .
          40 28 91 9   30 8   9   8

LESSON 2-7 Practice A
Multiplication Equations

Match each equation in Column 1 to its solution in Column 2.

Column 1
1. $5x = 40$ __F__
2. $21 = 3x$ __A__
3. $6x = 24$ __E__
4. $42 = 7x$ __J__
5. $2x = 18$ __D__
6. $20 = 4x$ __B__
7. $17x = 17$ __I__
8. $16 = 8x$ __H__
9. $20x = 0$ __G__
10. $27 = 9x$ __C__

Column 2
A. $x = 7$
B. $x = 5$
C. $x = 3$
D. $x = 9$
E. $x = 4$
F. $x = 8$
G. $x = 0$
H. $x = 2$
I. $x = 1$
J. $x = 6$

Solve each equation. Check your answers.

11. $6p = 30$
    $p = 5; 6 \cdot 5 = 30$

12. $4q = 12$
    $q = 3; 4 \cdot 3 = 12$

13. $21 = 3s$
    $s = 7; 21 = 3 \cdot 7$

14. $8 = 2w$
    $w = 4; 8 = 2 \cdot 4$

15. $25 = 5t$
    $t = 5; 25 = 5 \cdot 5$

16. $9m = 54$
    $m = 6; 9 \cdot 6 = 54$

17. Cheryl gets paid $8 per hour at her job at the record store. She made a total of $96 last week. Write a multiplication equation using the variable h to show how many hours she worked last week.
    $8h = 96$ h = 12 hours

18. There are 3 feet in a yard. John used 27 feet of wire in his sculpture. Write a multiplication equation using the variable y to find how many yards of wire John used in his sculpture.
    $3y = 27$; 9 yards

LESSON 2-7 Practice B
Multiplication Equations

Solve each equation. Check your answers.

1. $8s = 72$
   $s = 9; 8 \cdot 9 = 72$

2. $4v = 28$
   $v = 7; 4 \cdot 7 = 28$

3. $27 = 9q$
   $q = 3; 27 = 9 \cdot 3$

4. $12m = 60$
   $m = 5; 12 \cdot 5 = 60$

5. $48 = 6x$
   $x = 8; 48 = 6 \cdot 8$

6. $7n = 63$
   $n = 9; 7 \cdot 9 = 63$

7. $10t = 130$
   $t = 13; 10 \cdot 13 = 130$

8. $15p = 450$
   $p = 30; 15 \cdot 30 = 450$

9. $84 = 6v$
   $v = 14; 84 = 6 \cdot 14$

Solve each equation.

10. $49 = 7m$
    $m = 7$

11. $20r = 80$
    $r = 4$

12. $64 = 8x$
    $x = 8$

13. $36 = 4p$
    $p = 9$

14. $147 = 7d$
    $d = 21$

15. $11n = 110$
    $n = 10$

16. $12q = 144$
    $q = 12$

17. $25f = 125$
    $f = 5$

18. $128 = 16w$
    $w = 8$

19. A hot-air balloon flew at 10 miles per hour. Using the variable h, write and solve a multiplication equation to find how many hours the balloon traveled if it covered a distance of 70 miles.
    $10h = 70; h = 7$ hours

20. A passenger helicopter can travel 300 miles in the same time it takes a hot-air balloon to travel 20 miles. Using the variable s, write and solve a multiplication equation to find how many times faster the helicopter can travel than the hot air balloon.
    $20s = 300; s = 15$ times faster

LESSON 2-7 Practice C
Multiplication Equations

Solve each equation. Check your answers.

1. $24s = 144$
   $s = 6; 24 \cdot 6 = 144$

2. $5v = \frac{225}{15}$
   $v = 3; 5 \cdot 3 = \frac{225}{15}$

3. $4q = \frac{16}{2}$
   $q = 2; 4 \cdot 2 = \frac{16}{2}$

4. $(3^2)m = 45$
   $m = 5; 3^2 \cdot 5 = 45$

5. $266 = 38x$
   $x = 7; 266 = 38 \cdot 7$

6. $(4^2)n = 48$
   $n = 3; 4^2 \cdot 3 = 48$

7. $213t = 1,917$
   $t = 9; 213 \cdot 9 = 1,917$

8. $(15p) \cdot 4 = 660$
   $p = 11; (15 \cdot 11) \cdot 4 = 660$

9. $3v = 300$
   $v = 100; 3 \cdot 100 = 300$

Solve each equation.

10. $65m = 845$
    $m = 13$

11. $105r = 840$
    $r = 8$

12. $(2^4)x = 112$
    $x = 7$

13. $(10p) \cdot 21 = 1,890$
    $p = 9$

14. $42d = 210$
    $d = 5$

15. $36 = 3n \cdot 4$
    $n = 3$

16. $57 \cdot (4q) = 228$
    $q = 1$

17. $137t = 822$
    $t = 6$

18. $(3^3)w = 54$
    $w = 2$

19. If $xy = 56$, and $y - x = 1$, what are the values for x and y?
    $y = 8$ and $x = 7$

20. If $x = 4y$, and $x + y = 10$, what are the values for x and y?
    $x = 8$ and $y = 2$

LESSON 2-7 Reteach
Multiplication Equations

You can use tiles to help you solve multiplication equations.

Division undoes multiplication, so you can use division to solve multiplication equations.

To solve $3x = 12$, first use tiles to model the equation.

$3x = 12$

Next, divide each side of the equal sign into 3 equal groups

The number of tiles in one group represents the solution.

$x = 4$

Check: $3x = 12$
$3 \cdot 4 \stackrel{?}{=} 12$
$12 = 12$ ✓

Use tiles to solve each equation. Then check each answer.

1. $5x = 15$
   $x = 3$
2. $2x = 6$
   $x = 3$
3. $4x = 16$
   $x = 4$
4. $8x = 24$
   $x = 3$
5. $3x = 18$
   $x = 6$
6. $6x = 12$
   $x = 2$
7. $7x = 21$
   $x = 3$
8. $9x = 9$
   $x = 1$
9. $4x = 24$
   $x = 6$
10. $3x = 9$
    $x = 3$
11. $8x = 16$
    $x = 2$
12. $5x = 25$
    $x = 5$

---

81 Holt Mathematics

LESSON 2-7 Challenge: Regulation Sizes

Write a multiplication equation for the area of each regulation field or court. Then solve the equations to find the missing measurements. Remember: Area = length • width, or $A = l \cdot w$.

1. $l = 94$ ft, $w = ?$, $A = 4{,}700$ ft^2
   $4{,}700 = 94w;\ w = 50$
   What is the width of a regulation basketball court?
   50 feet

2. $l = ?$, $w = 75$ m, $A = 8{,}250$ m^2
   $8{,}250 = 75l;\ l = 110$
   What is the length of a regulation soccer field?
   110 meters

3. $l = ?$, $w = 26$ m, $A = 1{,}586$ m^2
   $1{,}586 = 26l;\ l = 61$
   What is the length of a regulation ice hockey rink?
   61 meters

4. $l = 90$ ft, $w = ?$, $A = 8{,}100$ ft^2
   $8{,}100 = 90w;\ w = 90$
   What is the width of a regulation baseball diamond?
   90 feet

LESSON 2-7 Problem Solving: Multiplication Equations

Write and solve a multiplication equation to answer each question.

1. In 1975, a person earning minimum wage made $80 for a 40-hour work week. What was the minimum wage per hour in 1975?
   $40x = 80;\ x = 2;$
   $2 per hour

2. If an ostrich could maintain its maximum speed for 5 hours, it could run 225 miles. How fast can an ostrich run?
   $5x = 225;\ x = 45;$
   45 miles per hour

3. About 2,000,000 people live in Paris, the capital of France. That is 80 times larger than the population of Paris, Texas. How many people live in Paris, Texas?
   $80x = 2{,}000{,}000;\ x = 25{,}000;$
   25,000 people

4. The average person in China goes to the movies 12 times per year. That is 3 times more than the average American goes to the movies. How many times per year does the average American go to the movies?
   $3x = 12;\ x = 4;$
   4 times per year

Circle the letter of the correct answer.

5. Recycling just 1 ton of paper saves 17 trees! If a city recycled enough paper to save 136 trees, how many tons of paper did it recycle?
   A 7 tons
   B 8 tons
   C 9 tons
   D 119 tons

6. Seaweed found along the coast of California, called giant kelp, grows up to 18 inches per day. If a giant kelp plant has grown 162 inches at this rate, for how many days has it been growing?
   F 180 days   H 9 days
   G 144 days   J 8 days

7. The distance between Atlanta, Georgia, and Denver, Colorado, is 1,398 miles. That is twice the distance between Atlanta and Detroit, Michigan. How many miles would you have to drive to get from Atlanta to Detroit?
   A 2,796 miles
   B 349.5 miles
   C 699 miles
   D 1,400 miles

8. Jupiter has 2 times more moons than Neptune has, and 8 times more moons than Mars has. Jupiter has 16 moons. How many moons do Neptune and Mars each have?
   F 8 moons, 2 moons
   G 2 moons, 8 moons
   H 128 moons, 32 moons
   J 32 moons, 128 moons

LESSON 2-7 Reading Strategies: Follow a Procedure

Multiplication and division are **inverse operations**. You can think of them as **opposite operations**.

$4 \cdot 12 = 48$ and $48 \div 12 = 4$
$6 \cdot 13 = 78$ and $78 \div 13 = 6$

From these examples, you could say that division **"undoes"** the multiplication.

Follow these steps to "undo" the multiplication and solve.

$7n = 84$ → Read: "7 times n equals 84."

Step 1: Get n by itself. Use division to "undo" multiplication. Since 7 is multiplied by n, divide by 7.
$7n = 84$

Step 2: To keep the equation balanced, divide the right side of the equation by 7 also.
$7n \div 7 = 84 \div 7$

Step 3: Check to verify that $n = 12$ is the solution.
$n = 12$
$7n = 84$
$7 \cdot 12 \stackrel{?}{=} 84$
$84 \stackrel{?}{=} 84$ ✓ 12 is the solution.

Answer each question.

1. What is another name for the "opposite operation"? **inverse operation**
2. What is the inverse operation for multiplication? **division**

Use $8z = 96$ for Exercises 3–6.

3. Write the equation in words. **8 times z equals 96.**
4. What operation is used in the equation? **multiplication**
5. What operation will you perform on both sides of the equation to solve it? **division**
6. Solve the equation. **$z = 12$**

LESSON 2-7 Puzzles, Twisters & Teasers: Go for the Gold!

Where did the 2002 Winter Olympics take place?
Find the letter that corresponds to the answer for each exercise using the decoder below. Then, place the letter in the blank corresponding to the exercise number to determine where the 2002 Winter Olympics were held.

1. At the 2002 Winter Olympics, Norway won 24 medals. Norway won 3 times as many medals as the Netherlands. How many medals did the Netherlands win?

   Solve: Check:
   $3n = 24$ $(3)(8) \stackrel{?}{=} 24$
   $\frac{3n}{3} = \frac{24}{3}$ $24 \stackrel{?}{=} 24$ ✓
   $n = 8$ medals

2. At the 2002 Games, the United States won 34 medals, a record number for the U.S. at any Winter Olympics. The U.S. won twice as many medals as Canada. How many medals did Canada win?

   Solve: Check:
   $34 = 2c$ $34 \stackrel{?}{=} (2)(17)$
   $\frac{34}{2} = \frac{2c}{2}$ $34 \stackrel{?}{=} 34$ ✓
   $17 = c$ medals

3. In short track speed skating, one lap is approximately 100 meters. About how many laps would a skater need to complete in order to finish a 1000 meter race?

   Solve: Check:
   $100x = 1000$ $(100)(10) \stackrel{?}{=} 1000$
   $\frac{100x}{100} = \frac{1000}{100}$ $1000 \stackrel{?}{=} 1000$ ✓
   $x = 10$ laps

4. In Cross-Country Skiing, the Combined Pursuit consists of 2 different ski styles. The freestyle portion for women is 10 km, twice as long as the classic style portion. How far must skiers race using the classic style?

   Solve: Check:
   $2c = 10$ $(2)(5) \stackrel{?}{=} 10$
   $\frac{2c}{2} = \frac{10}{2}$ $10 \stackrel{?}{=} 10$ ✓
   $c = 5$ km

A	B	D	E	G	H	I	K	L	M	N	O	R	S	T	U	W
10	12	7	21	6	5	18	24	13	14	19	9	22	11	17	8	23

U	T	A	H
1	2	3	4

LESSON 2-8 Practice A
Division Equations

Match each equation in Column 1 to its solution in Column 2.

Column 1		Column 2
1. $\frac{x}{4} = 5$	H	A. $x = 10$
2. $\frac{x}{3} = 8$	E	B. $x = 15$
3. $3 = \frac{x}{6}$	J	C. $x = 36$
4. $\frac{x}{7} = 7$	D	D. $x = 49$
5. $2 = \frac{x}{5}$	A	E. $x = 24$
6. $4 = \frac{x}{9}$	C	F. $x = 27$
7. $\frac{x}{1} = 5$	I	G. $x = 12$
8. $\frac{x}{3} = 5$	B	H. $x = 20$
9. $\frac{x}{3} = 9$	F	I. $x = 5$
10. $3 = \frac{x}{4}$	G	J. $x = 18$

Solve each equation. Check your answers.

11. $\frac{p}{4} = 4$
    $p = 16; \frac{16}{4} = 4$

12. $\frac{q}{8} = 3$
    $q = 24; \frac{24}{8} = 3$

13. $7 = \frac{s}{3}$
    $s = 21; 7 = \frac{21}{3}$

14. $4 = \frac{w}{9}$
    $w = 36; 4 = \frac{36}{9}$

15. $7 = \frac{t}{5}$
    $t = 35; 7 = \frac{35}{5}$

16. $\frac{m}{7} = 8$
    $m = 56; \frac{56}{7} = 8$

17. All of the students in Tim's class are divided into 4 teams of 6 students. Write a division equation using the variable s to show the total number of students in Tim's class.

$\frac{s}{4} = 6$

18. There are 3 tennis balls in each can. The coach bought a total of 27 tennis balls. Write and solve a division equation using the variable c to find how many cans the coach bought.

$\frac{27}{c} = 3; c = 9$ cans

/09

LESSON 2-8 Practice B
Division Equations

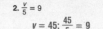

Solve each equation. Check your answers.

1. $\frac{s}{6} = 7$
   $s = 42; \frac{42}{6} = 7$

2. $\frac{v}{5} = 9$
   $v = 45; \frac{45}{5} = 9$

3. $12 = \frac{q}{7}$
   $q = 84; 12 = \frac{84}{7}$

4. $\frac{m}{2} = 16$
   $m = 32; \frac{32}{2} = 16$

5. $26 = \frac{x}{3}$
   $x = 78; 26 = \frac{78}{3}$

6. $\frac{n}{8} = 4$
   $n = 32; \frac{32}{8} = 4$

7. $\frac{t}{11} = 11$
   $t = 121; \frac{121}{11} = 11$

8. $\frac{p}{7} = 10$
   $p = 70; \frac{70}{7} = 10$

9. $7 = \frac{v}{8}$
   $v = 56; 7 = \frac{56}{8}$

Solve each equation.

10. $10 = \frac{m}{9}$
    $m = 90$

11. $\frac{r}{5} = 8$
    $r = 40$

12. $11 = \frac{x}{7}$
    $x = 77$

13. $9 = \frac{p}{12}$
    $p = 108$

14. $15 = \frac{d}{5}$
    $d = 75$

15. $\frac{n}{4} = 28$
    $n = 112$

16. $\frac{q}{2} = 134$
    $q = 268$

17. $\frac{u}{16} = 1$
    $u = 16$

18. $2 = \frac{w}{25}$
    $w = 50$

19. All the seats in the theater are divided into 6 groups. There are 35 seats in each group. Using the variable s, write and solve a division equation to find how many seats there are in the theater.

$\frac{s}{6} = 35; s = 210$ seats

20. There are 16 ounces in one pound. A box of nails weighs 4 pounds. Using the variable w, write and solve a division equation to find how many ounces the box weighs.

$\frac{w}{16} = 4; w = 64$ ounces

or $\frac{w}{4} = 16$

/22

LESSON 2-8 Practice C
Division Equations

Solve each equation. Check your answers.

1. $\frac{s}{18} = 16$
   $s = 288; \frac{288}{18} = 16$

2. $\frac{v}{24} = 3^2$
   $v = 216; \frac{216}{24} = 3^2$

3. $\frac{q}{9} = 2 \cdot 7$
   $q = 126; \frac{126}{9} = 2 \cdot 7$

4. $\frac{m}{2^2} = 35$
   $m = 140; \frac{140}{2^2} = 35$

5. $7 = \frac{x}{25}$
   $x = 175; 7 = \frac{175}{25}$

6. $\frac{n}{3^2} = 3 \cdot 4$
   $n = 108; \frac{108}{3^2} = 3 \cdot 4$

7. $\frac{t}{30} = 14$
   $t = 420; \frac{420}{30} = 14$

8. $\frac{p}{48} = 5$
   $p = 240; \frac{240}{48} = 5$

9. $\frac{v}{7} = 3^2$
   $v = 63; \frac{63}{7} = 3^2$

Solve each equation.

10. $\frac{m}{27} = 5$
    $m = 135$

11. $\frac{r}{9} = 41$
    $r = 369$

12. $\frac{x}{(2^3)} = 7$
    $x = 56$

13. $\frac{p}{(16 \cdot 2)} = 6$
    $p = 192$

14. $\frac{d}{59} = 2$
    $d = 118$

15. $9^2 = \frac{n}{3}$
    $n = 243$

16. $\frac{q}{35} = 11$
    $q = 385$

17. $\frac{t}{8} = \frac{117}{9}$
    $t = 104$

18. $4^3 = \frac{w}{2^3}$
    $w = 512$

19. If $\frac{x}{y} = 2$, and $x + y = 6$, what are the values of x and y?

$x = 4$ and $y = 2$

20. If $\frac{(3y)}{x} = 4$, and $x^2 = 36$, what are the values for x and y?

$x = 6$ and $y = 8$

LESSON 2-8 Reteach
Division Equations

You can use multiplication and division to write related number facts.

$3 \cdot 4 = 12$ $12 \div 4 = 3$

Division and multiplication are inverse operations. They undo each other. So you can use multiplication to solve division equations.

To solve $\frac{x}{2} = 3$, think of a related number fact.

If $\frac{x}{2} = 3$, then $3 \cdot 2 = x$.

$3 \cdot 2 = x$
$x = 6$

Check: $\frac{x}{2} = 3$
$\frac{6}{2} \stackrel{?}{=} 3$ substitute
$3 \stackrel{?}{=} 3$ ✓

$x = 6$ is the solution for $\frac{x}{2} = 3$.

Use a related number fact to solve each equation. Then check each answer.

1. $\frac{x}{2} = 4$
   $x = 8; \frac{8}{2} = 4$

2. $\frac{x}{8} = 2$
   $x = 16; \frac{16}{8} = 2$

3. $\frac{x}{3} = 5$
   $x = 15; \frac{15}{3} = 5$

4. $\frac{x}{5} = 1$
   $x = 5; \frac{5}{5} = 1$

5. $\frac{x}{9} = 3$
   $x = 27; \frac{27}{9} = 3$

6. $\frac{x}{6} = 3$
   $x = 18; \frac{18}{6} = 3$

7. $\frac{x}{8} = 4$
   $x = 32; \frac{32}{8} = 4$

8. $\frac{x}{2} = 9$
   $x = 18; \frac{18}{2} = 9$

9. $\frac{x}{4} = 4$
   $x = 16; \frac{16}{4} = 4$

10. $\frac{x}{5} = 4$
    $x = 20; \frac{20}{5} = 4$

11. $\frac{x}{6} = 2$
    $x = 12; \frac{12}{6} = 2$

12. $\frac{x}{9} = 4$
    $x = 36; \frac{36}{9} = 4$

/12

Copyright © by Holt, Rinehart and Winston.

83

Holt Mathematics

LESSON 2-8 Challenge
What Does Algebra Mean?

About 1,200 years ago, Arab people invented the branch of mathematics called *algebra*. In fact, the word *algebra* comes from the Arabic word *al-jabr*. What does that word mean?

Solve each division equation below. Then in the box at the bottom of the page, write the variable in the blank above its value. When you have solved all the equations you will have found the answer to the question.

1. $\frac{s}{4} = 6$ $s = 24$
2. $\frac{b}{3} = 5$ $b = 15$
3. $9 = \frac{p}{4}$ $p = 36$
4. $i \div 2 = 7$ $i = 14$
5. $8 = \frac{a}{6}$ $a = 48$
6. $11 = f \div 2$ $f = 22$
7. $\frac{h}{7} = 6$ $h = 42$
8. $\frac{k}{9} = 5$ $k = 45$
9. $6 = \frac{u}{2}$ $u = 12$
10. $3 = \frac{n}{7}$ $n = 21$
11. $o \div 8 = 4$ $o = 32$
12. $\frac{r}{7} = 5$ $r = 35$
13. $t \div 9 = 3$ $t = 27$
14. $8 = e \div 1$ $e = 8$

| 27 | 42 | 8 | 35 | 8 | 12 | 21 | 14 | 32 | 21 |

| 32 | 22 | 15 | 35 | 32 | 45 | 8 | 21 |

| 36 | 48 | 35 | 27 | 24 |

Answer: the reunion of broken parts

LESSON 2-8 Problem Solving
Division Equations

Use the table to write and solve a division equation to answer each question.

Bakersville Sports League		
Sport	Number of Teams	Players on Each Team
Baseball	7	20
Soccer	11	15
Football	8	24
Volleyball	12	9
Lacrosse	6	17
Basketball	10	10
Tennis	18	6

1. How many total people signed up to play soccer in Bakersville this year?
   $\frac{x}{11} = 15$; $x = 165$; **165 people**

2. How many people signed up to play lacrosse this year?
   $\frac{x}{6} = 17$; $x = 102$; **102 people**

3. What was the total number of people who signed up to play baseball this year?
   $\frac{x}{7} = 20$; $x = 140$; **140 people**

4. Which two sports in the league have the same number of people signed up to play this year? How many people are signed up to play each of those sports?
   volleyball and tennis; 108 people

Circle the letter of the correct answer.

5. Which sport has a higher total number of players, football or tennis? How many more players?
   A football; 10 players
   B tennis; 144 players
   (C) football; 84 players
   D tennis; 18 players

6. Only one sport this year has the same number of players on each team as its number of teams. Which sport is that?
   (F) basketball
   G football
   H soccer
   J tennis

LESSON 2-8 Reading Strategies
Follow a Sequence

Knowing that multiplication and division are **inverse operations** can help you solve division equations.

    $51 \div 3 = 17$ and $17 \cdot 3 = 51$
    $65 \div 5 = 13$ and $13 \cdot 5 = 65$

From these examples, you could say that multiplication **"undoes"** division. This makes sense, since multiplication and division are **opposite operations**.

Example: $s \div 18 = 5 \rightarrow$ Read: "*s* divided by 18 equals 5."

Follow these steps to solve:

Step 1: Get the variable by itself. Use multiplication to "undo" division. Since *s* is divided by 18, you will multiply by 18.
    $s \div 18 = 5$

Step 2: To keep the equation balanced, multiply the right side of the equation by 18 also.
    $s \div 18 \cdot 18 = 5 \cdot 18$

Step 3: Check to verify that $s = 90$ is the solution.
    $s = 90$
    $s \div 18 = 5$
    $90 \div 18 = 5$
    $5 \stackrel{?}{=} 5$ ✓   90 is the solution.

Use $w \div 7 = 98$ for Exercises 1–4.

1. Write the equation in words.
   w divided by 7 equals 98.

2. What operation is used in the equation?
   division

3. What operation will you perform on both sides of the equation to solve it?
   multiplication

4. Write about how you would solve this problem step by step: $7 = z \div 20$.
   possible steps: (1) Multiply both sides by 20; (2) Solve: $z = 140$; (3) Check: $\frac{140}{20} = 7$; 140 is the solution.

LESSON 2-8 Puzzles, Twisters & Teasers
Where Are They?

Where will you find a lobster's teeth?
Hint: It is called the gastric mill.

Solve each equation. Use the inverse operation to check your answers.

1. $\frac{x}{4} = 8$ Check:
   $4 \cdot \frac{x}{4} = 4 \cdot 8$ $\frac{32}{4} \stackrel{?}{=} 8$
   $x = 32$ $8 \stackrel{?}{=} 8$ ✓

2. $7 = \frac{y}{6}$ Check:
   $6 \cdot 7 = 6 \cdot \frac{y}{6}$ $7 \stackrel{?}{=} \frac{42}{6}$
   $42 = y$ $7 \stackrel{?}{=} 7$ ✓

3. $\frac{z}{9} = 5$ Check:
   $9 \cdot \frac{z}{9} = 9 \cdot 5$ $\frac{45}{9} \stackrel{?}{=} 5$
   $z = 45$ $5 \stackrel{?}{=} 5$ ✓

4. $64 = \frac{m}{2}$ Check:
   $2 \cdot 64 = 2 \cdot \frac{m}{2}$ $64 \stackrel{?}{=} \frac{128}{2}$
   $128 = m$ $64 \stackrel{?}{=} 64$ ✓

5. $\frac{s}{12} = 72$ Check:
   $12 \cdot \frac{s}{12} = 12 \cdot 72$ $\frac{864}{12} \stackrel{?}{=} 72$
   $s = 864$ $72 \stackrel{?}{=} 72$ ✓

6. $4 = \frac{n}{25}$ Check:
   $25 \cdot 4 = 25 \cdot \frac{n}{25}$ $4 \stackrel{?}{=} \frac{100}{25}$
   $100 = n$ $4 \stackrel{?}{=} 4$ ✓

7. $\frac{b}{10} = 100$ Check:
   $10 \cdot \frac{b}{10} = 10 \cdot 100$ $\frac{1000}{10} \stackrel{?}{=} 100$
   $b = 1000$ $100 \stackrel{?}{=} 100$ ✓

To find the answer, use your solutions in the decoder:

A	B	C	D	E	F	G	H	I	J	K	L	M	N
864	12	100	702	21	111	209	1000	180	243	13	128	19

O	P	Q	R	S	T	U	V	W	X	Y	Z
45	150	33	40	32	42	500	22	34	116	8	23

In it's   S   T   O   M   A   C   H
            1   2   3   4   5   6   7